JOURNAL OF
GREEN ENGINEERING

Volume 3, No. 3 (April 2013)

Special issue on

Emerging Green Energy Techniques

Guest Editor:

Johnson I. Agbinya

JOURNAL OF GREEN ENGINEERING

Chairperson: Ramjee Prasad, CTIF, Aalborg University, Denmark
Editor-in-Chief: Dina Simunic, University of Zagreb, Croatia

Editorial Board
Luis Kun, Homeland Security, National Defense University, i-College, USA
Dragan Boscovic, Motorola, USA
Panagiotis Demstichas, University of Piraeus, Greece
Afonso Ferreira, CNRS, France
Meir Goldman, Pi-Sheva Technology & Machines Ltd., Israel
Laurent Herault, CEA-LETI, MINATEC, France
Milan Dado, University of Zilina, Slovak Republic
Demetres Kouvatsos, University of Bradford, United Kingdom
Soulla Louca, University of Nicosia, Cyprus
Shingo Ohmori, CTIF-Japan, Japan
Doina Banciu, National Institute for Research and Development in Informatics, Romania
Hrvoje Domitrovic, University of Zagreb, Croatia
Reinhard Pfliegl, Austria Tech-Federal Agency for Technological Measures Ltd., Austria
Fernando Jose da Silva Velez, Universidade da Beira Interior, Portugal
Michel Israel, Medical University, Bulgaria
Sandro Rambaldi, Universita di Bologna, Italy
Debasis Bandyopadhyay, TCS, India

Aims and Scopes
Journal of Green Engineering will publish original, high quality, peer-reviewed research papers and review articles dealing with environmentally safe engineering including their systems. Paper submission is solicited on:

- Theoretical and numerical modeling of environmentally safe electrical engineering devices and systems.
- Simulation of performance of innovative energy supply systems including renewable energy systems, as well as energy harvesting systems.
- Modeling and optimization of human environmentally conscientiousness environment (especially related to electromagnetics and acoustics).
- Modeling and optimization of applications of engineering sciences and technology to medicine and biology.
- Advances in modeling including optimization, product modeling, fault detection and diagnostics, inverse models.
- Advances in software and systems interoperability, validation and calibration techniques. Simulation tools for sustainable environment (especially electromagnetic, and acoustic).
- Experiences on teaching environmentally safe engineering (including applications of engineering sciences and technology to medicine and biology).

All these topics may be addressed from a global scale to a microscopic scale, and for different phases during the life cycle.

JOURNAL OF GREEN ENGINEERING

Volume 3   No. 3   April 2013

Published, sold and distributed by:
River Publishers
P.O. Box 1657
Algade 42
9000 Aalborg
Denmark

Tel.: +45369953197
www.riverpublishers.com

Journal of Green Engineering is published four times a year.
Publication programme, 2012–2013: Volume 3 (4 issues)

ISSN 1904-4720

Editorial Foreword:
Emerging Green Energy Techniques

Johnson I. Agbinya

Department of Electronic Engineering, La Trobe University, Melbourne, Australia

Within the last decade green communications has continued to attract increasing attention across several areas including power line communications and smart grids, Internet of things, energy efficient sensor networks, cognitive radios, wireless power transfer, energy harvesting, green hardware and software, system reliability and maintenability and a host of other related areas. The objective in this Special Issue of the *Journal of Green Engineering* is to publish original papers from a recent conference with significant contribution to areas of green engineering. Of the seven papers presented in this issue of the Journal five of them originally appeared in more brief forms in the conference proceedings of IB2Com 2012, all dealing with realistic aspects of green communications and computing. The paper by Giblin et al. is included to complete the loop in the application areas of interest to green engineering.

In "Achievable data rates of broadband power line communications in an underground medium-voltage network" A. Waadt, C. Kocks, G.H. Bucks, P. Jung and B. Sachsenhauser provide detailed analysis of the achievable channel capacities (data rates) of new broadband power line (BPL) standards proposed by the European Union (EU) and compared their performances with existing published works. The authors have shown that "the theoretical channel capacity with the achievable data rates show that the data throughput could be improved, if the standardized BPL transmission techniques were tailored to cope with the extremely low SNR. However, even with improved transmission techniques, frequencies above 10 MHz cannot be used for appreciable data transmission in MV networks, if the EMC limits are not relaxed".

The applications of radio frequency identification (RFID) tags are widespread including tracking of objects, wireless power transfer and fleet management systems. However, privacy and security concerns remain two areas of interest in their applications. In "A one round-trip ultralight weight security protocol for low-cost RFID tags", W. Razouk and A. Sekkaki introduce a couple of novelties to enhance the low computational capability and limited storage capacity of low-cost RFID tags. They have proposed and introduced "the utilization of only two messages to fully complete the authentication and identification of the reader-tag". They also introduce "the implementation of the pseudo random number generator (PRNG) on the server side" which "reduces the storage and computation requirements on the tag". The "proposed scheme protects the user's privacy and resists several attacks like malicious traceability, replay and impersonation attacks". More importantly, the low computation procedure avoids the use of more expensive cryptographic schemes.

Judicious use of scarce resources in cognitive radio (CR) systems is currently one of the most exciting areas of research in wireless communications. M.A Shah, S. Zhang and C. Maple in the paper entitled "Efficient discovery and recovery of common control channel in cognitive radio wireless ad-hoc systems" have proposed a new protocol for searching, scanning and accessing of the control channel. Two steps are required consisting of rapid channel access and reliable channel access. The method helps nodes to quickly and efficiently converge to newly discovered control channels. This efficiency is essential for reduced power utilisation by the CR system.

The emerging convergence of wireless sensor networks, ad-hoc networks and wireless communication to the so-called "Internet of Things" means that more and more efficient energy utilisation schemes will be required to minimise the costs associated with powering of trillions of 'things' worldwide. D. Zrno, D. Šimunić and R. Prasad in the paper "Wireless sensor networks – Routing impact on energy distribution and energy hole formation" investigate the impact of different routing techniques on energy consumption in sensor networks with a central node. Sensor nodes closest to a central node tend to dedicate more resources to routing of packets to nodes far away thereby depleting their energy resources faster. This leads to energy holes. This leads to reduced life of the network and possibly network fragmentation. By testing the performance of several routing protocols, the authors conclude that the best performance is obtained with "optimal multi-hop routing which calculates the best direct path towards the target node and then optimizes it for multi-hop transmission".

The issue of sustainability is central to green engineering. This issue is tackled by A. Ito, Y. Hiramatsu, F. Shimada and F. Sato in their paper "Designing education process in an elementary school for mobile phone literacy" as a process to secure use of mobile phones in learning and social relationships. Steps outlined in the paper are essential for the design of green mobile phones for children of all ages and backgrounds.

There is no time in human history when disposable electronic waste is more of a problem than now. The culture of replacement rather than repair of faulty devices exacerbates the problem leaving maintainable hardware to be disposed of instead of having them repaired. K. Aboura and J.I. Agbinya present "a procedure for determining optimal times for the replacement of a large number of identical items operating under similar conditions. A collective maintenance policy is considered due to the prohibitive cost of individual replacement upon failure". They proposed that the first replacement time should be chosen using initial reliability estimates for the items under consideration. Their approach could help to reduce for example mass replacement of computing devices by organisation all at once instead of progressive replacement.

The last paper "Heart rate variability, blood pressure and cognitive function: Assessing age effects" by L. Giblin, L. De Leon, L. Smith, T. Sztynda and S. Lal provides an application area for green engineering in this case monitoring of bio-conditions such as heart rate which has the potential to reduce illnesses before they occur and their consequences on resource utilisation.

About the Editor

Johnson I. Agbinya is currently Associate Professor in the Department of Electronic engineering at La Trobe University, Melbourne, Australia. He is also Honorary Professor at the University of Witwatersrand (WITS), South Africa; Extraordinary Professor at the University of the Western Cape (UWC), Cape Town and the Tshwane University of Technology (TUT), Pretoria, South Africa. Prior to joining La Trobe University in November 2011, he was Senior Research Scientist at CSIRO Telecommunications and Industrial Physics (now CSIRO ICT) from 1993–2000, Principal Research Engineering at Vodafone Australia (2000–2003) and Senior Lecturer at UTS Australia (2003–2011). His R&D activities cover remote sensing, Internet of things (machine to machine communications), bio-monitoring systems, wireless power transfer, mobile communications and biometrics systems. He has authored/co-author nine technical books in telecommunications, some of which are used as textbooks. He is founder of the International Conference on Broadband Communications and Biomedical Applications (IB2COM), Pan African Conference on Science, Computing and Telecommunications (PACT) and the *African Journal of Information and Communication Technology* (AJICT). He has published more than 250 peer-reviewed research publications in international journals and conference proceedings. He received his BSc degree electronic/electrical engineering from Obafemi Awolowo University (OAU), Ile Ife, Nigeria; MSc in electronic engineering from the University of Strathclyde, Glasgow, Scotland and PhD from La Trobe University, Melbourne, Australia in 1973, 1982 and 1994 respectively. He received Best Paper award from IEEE 5th International Conference on Networking (ICN'2006) Mauritius, CSIRO ADCOM group research award in 1997 and Research Trailblazer Certificate at UTS in 2009. He is the Editor in Chief of the *African Journal of ICT* (AJICT), General Chair of several international conferences and member of several current international technical conference committees. He has served as expert on several international grants reviews/committees and was a rated researcher by the South African National Research Fund (NRF).

Achievable Data Rates of Broadband Power Line Communications in an Underground Medium-Voltage Network

Andreas Waadt[1], Christian Kocks[1], Guido H. Bruck[1], Peter Jung[1] and Bernd Sachsenhauser[2]

[1]Department of Communication Technologies, University of Duisburg-Essen, 47057 Duisburg, Germany; e-mail: info@kommunikationstechnik.org
[2]Division Smart Grid, Siemens AG, Germany

Received 30 January 2013; Accepted 27 March 2013

Abstract

Power line communications (PLC) or broadband over power lines (BPL) are getting popular for communication in smart grids. Consequently, new BPL standards have been introduced and new norms for the electromagnetic compatibility (EMC) of BPL are approaching. In this paper, the gross data rate is evaluated, which becomes achievable, if the newly approaching European EMC norm prEN 50561 is applied to IEEE 1901 compliant BPL. For comparison also older EMC norms are considered. The achievable data rate depends on the standardized transmission techniques and on the signal-to-noise ratio (SNR). To evaluate the SNR in underground medium-voltage (MV) networks, the channel attenuation and the noise have been measured on a 700 m MV line and compared to previous publications.

Keywords: Broadband over power lines, channel capacity, channel measurements, medium-voltage networks, power line communication.

Journal of Green Engineering, Vol. 3, 245–259.

1 Introduction

Next generation smart grids require a multiplicity of enhanced communication technologies [1]. Due to the direct connection with the electricity meters, power line communications (PLC) or broadband over power lines (BPL) become the first choice for communication in smart grids [2]. A measure of the performance and value of a communication system is the channel capacity. For BPL in access networks it depends on:

- the channel characteristics,
- the characteristics of noise and disturbances,
- the transmission techniques,
- the coupling of the signal to the power lines and
- the transmission power and, thus, on the applicable electromagnetic compatibility (EMC) norms.

The characteristics of channel and disturbances for BPL in low-voltage (LV) grids are well researched [3–6]. The characteristics of narrowband power line communications (PLC) are thoroughly analyzed as well [7, 8]. In recent years also the characteristics and capacity of broadband channels in medium-voltage (MV) access networks have been investigated [9–12]. Based on measurements, transmission channel and noise in MV networks have been characterized in [9] and [10]. Also the channel capacity has been calculated in [10]. The reachable capacity depends on the injected power spectral density (IPSD). In literature, the IPSD was determined by assuming a certain total transmission power P_{tx} and applying algorithms such as the water-filling (W-F) approach to optimize the spectral efficiency. In [10] a total transmission power of $P_{tx} = 1$ W was assumed, yielding channel capacities of tens or hundreds of Mbit/s. However, with this approach, the limits of the electromagnetic compatibility (EMC) are exceeded by tens of dB. EMC limits were taken into account in [11] and [12], where also channel capacities were determined. However, in these cases, the channel attenuation, used in the capacity calculations, were based on calculations of theoretical overhead MV networks, which were specified by the network geometry in terms of cable lengths and number of branches between transmitter and receiver. A comparison with measurement results in underground MV networks show significant differences of the attenuation [10]. Independently on the absolute attenuation, it was shown that channel capacity is very sensitive to EMC limits [11, 12]. The question, which EMC norm shall be applied to BPL, is a matter of dispute. The European norm EN 55022 and the newly emerging draft prEN 50561, which is specifically tailored to BPL were not yet considered for the calculations of the channel capacities. Furthermore, the channel capacity is only a theoretical upper threshold of the maximally possible data rate. Actually, the achievable data rate also depends on transmission techniques. Those transmission techniques to be used in BPL were recently standardized in IEEE 1901 [13].

In this paper, new measurement results from an underground MV network are presented, cf. Section 2. The results are similar to measurement results, which were found in other underground MV networks [10]. The found channel characteristics are used to determined channel capacities, taking the EMC limits from EN 55022 and the newly approaching draft prEN 50561 into account, cf. Section 3. To evaluate more realistic boarders of achievable data rates, block error ratios (BLER) have been simulated for transmission techniques from the BPL standard IEEE 1901, cf. Section 4. From IEEE 1901 only the physical layer (PHY) based on Wavelet OFDM is simulated with the concatenation of Reed Solomon (RS) and convolutional codes (CC) used as forward error correction (FEC) scheme. The minimally required signal-to-noise ratios (SNR) are identified from the simulation results and used to calculate the usable bandwidth and finally the achievable data rates, cf. Section 5. A conclusion is given in Section 6.

2 Measurements

2.1 Measurement Environment and Setup

For the characterization of the BPL transmission channel, measurements were carried out at premises of a 10 kV underground medium-voltage (MV) access network in a city in Germany. The magnitude of the channel's transfer function and the noise were measured at two locations of the MV network, a distribution substation S_D and a transformer substation S_T. Figure 1 shows the measurement setup in the access network. S_D and S_T are connected by a 700 m underground MV cable. Capacitive PLC couplers C_{C1} and C_{C2} are used to connect the measurement equipment in S_D and S_T in common mode to one lead of the MV power line. The insertion loss is less than 2 dB. Better transmission characteristics could be achieved by coupling the signal differentially between two leads, as it is done in LV networks. But such differential coupling requires a second capacitive coupler and PLC couplers for MV are expensive. Thus, in practice, the signal is often coupled in common mode. Except of C_{C1}, C_{C2} and the fuse, there are no other elements between transmitter and receiver. S_D, S_T and the MV line have connections to other distribution

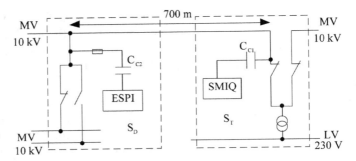

Figure 1 Elements of the measured MV access networks.

and transformer substations. The transformer of S_T is also connected to a 230 V low-voltage (LV) network of an urban district.

2.2 Transfer Function

A Rohde & Schwarz (R&S) signal generator (SMIQ) is used to transmit monochromatic signals at different frequencies f_i at the transformer substation S_T, while a R&S spectrum analyzer and test receiver (ESPI) is used at the distribution substation S_D, measuring the received signal power in the frequency range from $f_i - B_r/2$ to $f_i + B_r/2$, with $B_r = 3$ kHz being the measurement bandwidth. With the powers $P_{rx,i}$ of the received signal and $P_{tx,i}$ of the transmitted signal during the i-th measurement, the transfer function's magnitude at frequency f_i becomes $|H(f_i)|^2 = P_{rx,i}/P_{tx,i}$. To get an impression of the frequency selective transfer function, the measurements have been repeated for different frequencies f_i, 0.5 MHz $\leq f_i \leq 20$ MHz, in steps of $f_{i+1} - f_i = 250$ kHz. Figure 2 shows the measurement results.

   A comparison to previous publications on the transfer function's magnitude or the end-to-end attenuation shows some differences. The measurement results in an underground MV network in Beijing showed similar results but slightly higher attenuations at low frequencies and slightly less attenuation at high frequencies [10]. However, totally different results of significantly lower attenuations at high frequencies were published in [12], where the attenuation was calculated for overhead MV networks based on theoretical considerations. Measurements in LV networks had shown attenuations, which look qualitatively similar to those from Figure 2, but with higher attenuations [14]. The measurement in Figure 2 has been repeated two

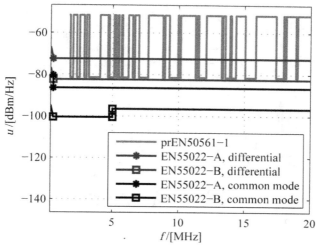

Figure 4 Spectral masks for mean conducted voltage in CISPR 22/EN 55022 and prEN 50561.

For the capacity calculation, the transmission power spectral density $P_{tx}(f)$ shall be as high as possible, i.e. equal to the EMC limits. Figure 4 shows the spectral masks from different European norms. Actually, the norms specify the maximally injected voltage in dBμV, to be measured on a reference bandwidth of 9 kHz. Figure 4 shows the power spectral density in dBm/Hz at a reference impedance $R = 50 \, \Omega$, which is the impedance of the PLC couplers.

The EN 55022, a copy of the international CISPR 22 norm from the Comit International Spcial des Perturbations Radiolectriques (CISPR), defines limits for two classes of devices, class A and class B, and for two different coupling modes, differential mode and common mode [17]. Differential mode means the voltage between two leads, whereas common mode means the voltage between lead and ground. The common mode limits are lower than the differential mode limits, because common mode voltages yield more radiation and interference to wireless services. The EN 55022 is actually not tailored to BPL but to electrical devices in general, its applicability to BPL is controversial. The new draft norm prEN 50561 is tailored to BPL and might be used for BPL in future [18–20]. It defines a relatively liberal limit of 95 dBμV/9 kHz and notches at operation frequencies of wireless services. To protect the wireless services from interference, the signal voltage shall be reduced by another 30 dB inside the notches. Figure 5 shows the resulting

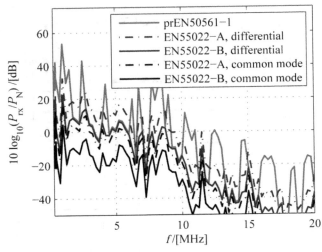

Figure 5   Reachable signal-to-noise ratio (SNR) on a 700 m MV line.

Table 1   Channel capacity over a 700 m MV line with different EMC limits.

f/MHz	EMC norm	C/[Mbit/s]
1.6065–20	prEN 50561-1	32.6
	EN 55022 Class A, differential	27.6
	EN 55022 Class B, differential	10.2
	EN 55022 Class A, common mode	5.90
	EN 55022 Class B, common mode	0.59
0.5–20	EN 55022 Class A, differential	32.7
	EN 55022 Class B, common mode	1.11
0.5–10	EN 55022 Class A, differential	32.4
	EN 55022 Class B, common mode	1.11

SNR at the receiver, if the channel from Figure 2 and the noise at S_D from Figure 3 are assumed.

Table 1 shows the resulting channel capacity for different frequency ranges and with different transmission powers according to the introduced EMC limits. A capacity of up to 32.7 Mbit/s could be reached. It shall be noted that the prEN 50561 limits are not defined below 1.6065 MHz. However, higher data rates can be achieved, if transmission starts at 0.5 MHz. On the other hand, due to the high attenuations at high frequencies, signal transmissions above 10 MHz is not worth it.

4 Required Signal-to-Noise Ratio

The channel capacities in Table 1 are upper thresholds for data rates. In practice, the achievable data rate depends not only on the SNR but also on the transmission techniques and the bandwidth, which is gainfully usable. Transmission techniques for BPL are standardized in IEEE 1901 [13]. To figure out, which SNR IEEE 1901 devices require, Monte Carlo simulations have been carried out. With the minimally required SNR and the frequency dependent SNR from Figure 5 the bandwidth is determined, which can be used for BPL transmission. Finally, with the applied modulation and coding scheme, the achievable gross data rate is calculated.

IEEE 1901 includes two physical layer (PHY) specifications, one using fast Fourier transform (FFT) orthogonal frequency division multiplexing (OFDM) and one using Wavelet OFDM. Both PHYs support different modulation and coding schemes (MCS) with different bit loadings and code rates. The channel codes used with the FFT OFDM PHY are turbo convolutional codes, whereas the Wavelet OFDM PHY uses either low-density parity-check convolutional codes (LDPC-CC) or concatenated Reed–Solomon and convolutional codes. The Wavelet OFDM modulation has advantages compared to FFT OFDM in BPL [21]. In the following, only the Wavelet OFDM PHY is considered for the transmission with concatenated Reed–Solomon and convolutional codes.

The minimally required SNR depends on the bit error ratios (BER) and the block error ratio (BLER), which are maximally tolerated by the service. Considering the transmission techniques and the position of BPL in the communication system, the BPL interface becomes comparable with a wireless interface in a communication network. Requirements for BER and BLER in wireless interfaces are defined in several standards and have been summarized in [22]. Several services with different quality-of-service (QoS) constraints are distinguished. Typically, BPL supports packet services using internet protocol (IP) to allow a plurality of applications. A BLER requirement for packet services with standard priority is BLER $< 10^{-4}$ [22]. Table 2 shows the found requirement for the ratio of bit energy E_b to noise spectral density N_0 for the introduced transmission technique from [13] and for different code rates $R_c = 1/2$ and $R_c = 7/8$.

The relation between bit energy to noise spectral density ratio $10 \log_{10}(E_b/N_0)$ and the SNR from Figure 5 depends on the modulation order M, the code rate R_c and the ratio of sampling rate and bandwidth. If

Table 2 Minimally required bit energy to noise ratio and signal power to noise ratio.

BLER	M	R_c	$10\log_{10}(E_b/N_0)$	$10\log_{10}(P_{rx}/P_N)$
10^{-4}	2	1/2	3.3 dB	0.29 dB
10^{-4}	2	7/8	6.0 dB	5.42 dB

Table 3 Usable bandwidth and data rates.

$f/$[MHz]	EMC norm	R_c	$B_u/$[MHz]	$R/$[Mbit/s]
1.6065–20	prEN50561	1/2	6.75	2.28
		7/8	4.25	2.52
1.6065–20	CISPR 22 A diff.	1/2	6.50	2.20
		7/8	5.75	3.41
1.6065–20	CISPR 22 B diff.	1/2	3.75	1.27
		7/8	1.50	0.89
1.6065–20	CISPR 22 A com.	1/2	2.25	0.76
		7/8	0.50	0.30
1.6065–20	CISPR 22 B com.	1/2	0.00	0.00
		7/8	0.00	0.00
0.5–10	CISPR 22 A diff.	1/2	7.38	2.50
		7/8	6.63	3.93
0.5–10	CISPR 22 B com.	1/2	0.25	0.08
		7/8	0.00	0.00

sampling rate and bandwidth are the same, the SNR becomes

$$10\log_{10}(P_{rx}/P_N) = 10\log_{10}(E_b/N_0) + 10\log_{10}(R_c) + 10\log_{10}(\log_2 M). \tag{3}$$

The Wavelet OFDM PHY in IEEE 1901 uses 512 carriers, occupying frequencies from 0 Hz to 31.25 MHz. This yields a carrier spacing $B_{SC} \approx$ 61 kHz. Optionally, only 1/2 or 1/4 of the full bandwidth can be occupied, yielding 15.625 MHz or 7.8125 MHz, respectively. However, due to the low achievable SNR from Figure 5 and the SNR constraints from Table 2, only a few MHz can be used for data transmission. Table 3 shows the remaining usable bandwidth B_u for different code rates R_c, EMC norms and frequency ranges.

5 Achievable Data Rates

The achievable gross data rate is calculated from the remaining usable bandwidth B_u, the code rate R_c, the bit loading ld(M) of the modulation order M

and the overhead of the guard intervals (GIs). In multi-carrier systems GIs are often added between the symbols to avoid inter symbol interference (ISI). FFT OFDM often uses cyclic GIs, referred to as cyclic prefix, which allow one-tap equalization in frequency domain. A similar approach is used in case of Circular Wavelet OFDM [23]. IEEE 1901 proposes GIs of 756 samples or 1956 samples, which can be used optionally in MV networks where channel impulse responses are longer, cf. section 13.4.2, PPDU structure, in [13]. The FFT size shall be 4096 points or 512 points for coexistence with TIA-1113 systems like HomePlug, cf. Section 13.2, FFT system, in [13]. With the 4096 FFT and 1956 samples GI for MV networks, the gross data rate becomes

$$R = \mathrm{ld}(M) \cdot R_\mathrm{c} \cdot B_\mathrm{u} \cdot \left(1 - \frac{1956}{4096 + 1956}\right). \tag{4}$$

The last column in Table 3 shows the achievable gross data rate from (4) for different EMC norms. In practice, the data rate is further reduced by protocol overhead and multiple access protocols.

The remaining achievable gross data rate is between about 80 kbit/s, if CISPR 22 limits for class B devices with common mode coupling are applied, and about 3.9 Mbit/s for CISPR 22 class A devices with differential mode coupling. The limits from the newly approaching prEN 50561 norm allow data rates of 2.5 Mbit/s. This result is far below estimations in other publications [10, 12]. Main reasons for this result are the combination of the high attenuation of the underground MV line with the low European EMC limits, which were not considered yet, and the transmission techniques from IEEE 1901, which are not tailored to cope with the resulting low SNR.

6 Conclusion

Measurements of channel attenuations and noise in an underground MV network have been carried out. In comparison with previous publications the measurements show similar results but slightly higher attenuations and a lower noise level. Taking different European EMC norms into account, the achievable SNR has been evaluated. With the in IEEE 1901 BPL minimally required SNR, the usable bandwidth and finally the achievable gross data rates have been determined, yielding only a few Mbit/s. Also the channel capacity has been calculated and compared to previous publications. Previous investigations determined higher capacities, either because no EMC limits were taken into account, or because higher EMC limits and lower, theoretically calculated channel attenuations were considered. A comparison of the theoretical channel capacity with the achievable data rates show that the data throughput could be improved, if the standardized BPL transmission techniques were tailored to cope with the extremely low SNR. However, even with improved transmission techniques, frequencies above 10 MHz cannot be used for appreciable data transmission in MV networks, if the EMC limits are not relaxed.

Acknowledgement

The measurements and characterization of the MV channel were part of a cooperation project, supported by the Siemens AG, division Smart Grid. The authors would like to thank for their continuous support and fruitful discussions throughout this research.

References

[1] V. C. Güngör, D. Sahin, T. Kocak, S. Ergüt, and C. Buccella. Smart grid technologies: Communication technologies and standards. IEEE Transactions on Industrial Informatics, 7:529–539, 2011.

[2] R. P. Lewis, D. P. Igic, and D. Z. Zhou. Assessment of communication methods for smart electricity metering in the U.K. In Proceedings IEEE PES/IAS Conference on Sustainable Alternative Energy (SAE), 2009.

[3] M. Zimmermann. Energieverteilnetze als Zugangsmedium für Telekommunikationsdienste. Berichte aus der Kommunikationstechnik, Shaker, 2000.

[4] A. Schwager, L. Stadelmeier, and M. Zumkeller. Potential of broadband power line home networking. In Proceedings Second IEEE Consumer Communications and Networking Conference (CCNC), 2005.

[5] Open PLC European Research Alliance, http://www.ist-opera.org/.

[6] H. C. Ferreira, L. Lampe, J. Newbury, and T. G. Swart Power Line Communications: Theory and Applications for Narrowband and Broadband Communications over Power Lines. John Wiley & Sons, Chichester, United Kingdom, 2010.

[7] M. Katayama, T. Yamazato, and H. Okada. A mathematical model of noise in narrowband power line communication systems. IEEE Journal on Selected Areas in Communications, 24:1267–1276, 2006.

[8] B. Varadarajan, I. H. Kim, A. Dabak, D. Rieken, and G. Gregg. Empirical measurements of the low-frequency power-line communications channel in rural North America. In Proceedings IEEE International Symposium on Power Line Communications and Its Applications (ISPLC2011), pp. 463–467, April 2011.

[9] J. J. Lee, S. J. Choi, H. M. Oh, W. T. Lee, K. H. Kim, and D. Y. Lee. Measurements of the communications environment in the medium voltage power distribution lines for wide-band power line communications. In Proceedings IEEE International Symposium of Power Line Communications and Its Applications (ISPLC04), Zaragoza, Spain, pp. 69–74, March 2004.

[10] H. Liu, J. Song, B. Zhao, and X. Li. Channel study for medium-voltage power network, in Proceedings IEEE International Symposium on Power Line Communications and Its Applications, pp. 245–250, 2006.

[11] P. Amirshahi and M. Kavehrad. High-frequency characteristics of overhead multiconductor power lines for broadband communications. IEEE Journal on Selected Areas in Communications, 24:1292–1303, July 2006.

[12] A. Lazaropoulos and P. Cottis. Capacity of overhead medium voltage power line communication channels. IEEE Transactions on Power Delivery, 25:723–733, April 2010.

[13] IEEE Standard for Broadband over Power Line Networks: Medium Access Control and Physical Layer Specifications, 2010.

[14] M. Gebhardt, F. Weinmann, and K. Dostert. Physical and regulatory constraints for communication over the power supply grid, *IEEE Communications Magazine*, 41:84–90, May 2003.

[15] D. Hansen. Review of EMC main aspects in fast PLC including some history. In Proceedings IEEE International Symposium on Electromagnetic Compatibility (EMC'03), vol. 1, Istanbul, Turkey, pp. 184–192, May 2003.

[16] R. Gallager. Information Theory and Reliable Communication. Wiley, 1968.

[17] Information technology equipment – Radio disturbance characteristics – Limits and methods of measurement, 2006.

[18] CLC/TC 210/WG 11, PLT apparatus standard.

[19] Powerline communication apparatus used in low voltage installations – Radio disturbance characteristics – Limits and methods of measurement – Part 1: Apparatus for in-home use, final draft, February 2012.

[20] Power-line telecommunications modems – Radio disturbance characteristics – Limits and methods of measurement – Part 2: Modems for access networks, prEN 50561-2:2011, 2011.

[21] S. Galli, H. Koga, and N. Kodama. Advanced signal processing for PLCs: Wavelet-OFDM. In *Proceedings IEEE International Symposium on Power Line Communications and Its Applications (ISPLC2008)*, Jeju City, Island, 2008.

[22] L. K. Tee. Packet error rate and latency requirements for a mobile wireless access system in an IP network. In Proceedings IEEE 66th Vehicular Technology Conference (VTC2007), Baltimore, USA, October 2007.

[23] L. Zbydniewski and T. Zielinski. Performance of wavelet-OFDM and circular wavelet-OFDM in power line communications. In Proceedings Signal Processing Algorithms, Architectures, Arrangements, and Applications (SPA2008), Poznan, Poland, September 2008.

Biographies

Andreas Waadt received the diploma (M.Sc. equiv.) in electrical engineering from the University of Duisburg-Essen, Germany, in 2004. Since 2004 he has been with the Department of Communication Technologies (Lehrstuhl für Kommunikationstechnik) at the University of Duisburg-Essen. His research areas include wireless communications, location tracking in mobile networks, cognitive radio and broadband over power lines.

Christian Kocks received the diploma (M.Sc. equiv.) in electrical engineering from the University of Duisburg-Essen, Germany, in 2008. He has been with the Department of Communication Technologies (Lehrstuhl fr Kommunikationstechnik) at the University of Duisburg-Essen since 2008 where he focuses on software-defined radio, cognitive radio and broadband powerline communications.

Guido H. Bruck has been with the Faculty of Electrical Engineering of the Gerhard-Mercator-Universität Duisburg since 1984. He joined the Department Communication Equipment and Systems (Nachrichtengeräte und anlagen) since then and worked in the field of source image coding. He developed a method to improve the quality of source coded images which contain high saturated colours. This can be done by considering the gamma distortion and compensation, which can be found in nearly all common image transmission systems. He adapted this method to image transmission systems like JPEG (Joint photographic experts group) and MPEG (Moving pictures experts group). The image quality can be improved compared to a standard JPEG or MPEG encoding or the amount of encoded data can be reduced by having the same image quality compared to a standard JPEG or MPEG encoding, if the image contains areas with high saturated colours. When Prof. Dr.-Ing. habil. Peter Jung joined the faculty in June 2000 the name of the department changed to Communication Technologies (KommunikationsTechnik). Since then, Dr. Bruck has worked in the field of software defined radio and on adaptation of source coding methods to mobile communication systems. Dr. Bruck is now Akademischer Oberrat (Senior Member of the Staff) at KommunikationsTechnik, being responsible for the administration of KommunikationsTechnik and managing several industrial projects and subprojects.

Peter Jung received the diploma (M.Sc. equiv.) in physics from University of Kaiserslautern, Germany, in 1990, and the Dr.-Ing. (Ph.D.EE equiv.) and Dr.-Ing. habil. (D.Sc.EE equiv.), both in electrical engineering with focus on microelectronics and communications technology, from University of Kaiserslautern in 1993 and 1996, respectively. In 1996, he became private educator (equiv. to reader) at University of Kaiserslautern and in 1998 also at Technical University of Dresden, Germany. From March 1998 till May 2000 he was with Siemens AG, Bereich Halbleiter, now Infineon Technologies, as Director of Cellular Innovation and later Senior Director of Concept Engineering Wireless Baseband. In June 2000, he became Chaired Professor for Communication Technologies (KommunikationsTechnik) at the Gerhard-Mercator-University Duisburg. In 1995, he was co-recipient of the best paper award at the ITG-Fachtagung Mobile Kommunikation, Ulm, Germany, and in 1997, he was co-recipient of the Johann-Philipp-Reis-Award for his work on multicarrier CDMA mobile radio systems. Professor Jung served as chairman of the Fakultätentag für Elektrotechnik und Informationstechnik (FTEI) e.V., and member of the board of VDE/VDI-GMM. He was a member of the editorial board of *IEEE Transactions on Wireless Communications* and has been a member of the editorial boards of the Springer journal *Wireless Personal Communications* and of *Hindawi Research Letters in Communications*. His areas of interest include wireless communication technology, software defined radio, and system-on-a-chip integration of communication systems.

Bernd Sachsenhauser graduated as Associate Engineer in Information and Computer Systems on the Berufsakademie of Siemens in Munich, Germany, in 1996. In May 2009 he received the Master of Business Administration (MBA) from the Henley Business School/University of Reading in England. From August 1996 till July 2006 he was with Siemens AG, Communication Segment, now Nokia Siemens Networks as Trainer for Communication Technologies. From August 2006 till August 2011 he joined the Energy Segment of Siemens with different functions in project and product responsibilities. Since October 2011 he is located in the Infrastructure and Cities Segment of Siemens and is responsible for Smart Grid Communication Projects and Solutions.

A One Round-Trip Ultralightweight Security Protocol for Low-Cost RFID Tags

Wissam Razouk and Abderrahim Sekkaki

Mathematics and Computer Science Department, Faculty of Science, Hassan II University, Maarif, B.P. 5366, 20000 Casablanca, Morocco; e-mail: {wissam.razouk, abderrahim.sekkaki}@etude.univcasa.ma

Received 15 February 2013; Accepted 27 March 2013

Abstract

Radio Frequency Identification systems are already used in many sensitive areas, and are likely to be adopted almost everywhere. But their massive deployment implies several security and privacy issues. Moreover, low cost RFID tags have very limited storage and computational capabilities and cannot afford classic cryptographic primitives, which makes them vulnerable to several attacks. In this paper, we introduce some novelties in the field of low-cost security protocols such as the utilization of only two messages to fully complete the authentication and identification of the reader-tag. Additionally, the implementation of the pseudo random number generator (PRNG) on the server side reduces the storage and computation requirements on the tag. Furthermore, the proposed scheme protects the user's privacy and resists several attacks like malicious traceability, replay and impersonation attacks. And most importantly our protocol relies on simple bitwise operations and does not require computationally expensive cryptographic mechanisms.

Keywords: RFID security, low-cost RFID, lightweight cryptography, mutual authentication, un-traceability, privacy, confidentiality.

Journal of Green Engineering, Vol. 3, 261–272.

1 Introduction

RFID tags are wireless microchips that help to identify objects and people as well. They are already used nowadays in a variety of applications like medical implants, public transportation, aviation security and homeland security, etc. [27]. They are also significantly popular in areas such as authentication, e-payment, access control, and more importantly in supply chain management, retail inventory control and product tracking [24]. This success is mainly due to the fact that RFID tags allow a very effective remote identification of a great number of tagged items simultaneously and without visual or physical contact.

Usually the tags and readers are the main components of a RFID system. The reader is generally connected to a database on a server that stores the information associated with the tags. However, the shared radio medium allows easy eavesdropping on the communication between the tag and the reader. Furthermore, the tag can also be queried without authorization. Indeed, although security is essential for many RFID systems, numerous applications may suffer from multiple vulnerabilities like information leakage, traceability, impersonation or denial of service.

In general, RFID tags can be categorized into different types, depending on power source, memory size or tag price. If we take the price as a criterion, we can divide RFID tags into high-cost and low-cost RFID tags. We will focus in the rest of this document on the last type, as for massive deployment the price has to be kept low. Indeed, due to market consideration, low-cost RFIDs are estimated to be the best choice, and are more likely to be widely deployed.

Unfortunately low-cost RFID tags are very constrained devices and lack resources for performing classic cryptographic primitives. These systems can store no more than hundreds of bits and have only 5–10 K logic gates [18]. Moreover, only 250-3 K can be used for security reasons.

Many proposals have been published for this purpose; nevertheless, obtaining a maximum security with very few resources is still considered as a real challenge [3].

In this paper we propose an efficient ultralightweight security protocol for low-cost RFID tags. Furthermore, the proposed scheme resists several attacks, provides mutual authentication and protects the user's privacy.

The rest of this paper is organized as follows: First, we give a detailed description of the proposed scheme in Section 4. Next, we present the security analysis and performance evaluation in Section 5. Finally, conclusions are given in Section 6.

2 Related Work

In 2006 three ultralightweight security protocols called UMAP family were proposed; Peris-Lopez et al. presented first M2AP [18], followed by EMAP [22] and LMAP [19]. Although considered as an interesting advance in lightweight cryptography, many security issues were highlighted using active and passive attacks [2, 4, 5, 8, 15–17].

Then in 2007 Chien released a proposal named SASI (strong authentication and strong integrity) [7]. The protocol shares the same structure with the UMAP family; it was designed to avoid some of the previous issues. But several security analyses have demonstrated de-synchronization attacks, and many weaknesses have been published [6, 9, 10, 13, 25].

Next, in 2008 Gossamer [21] was proposed to further develop the UMAP family and SASI protocol; however, other papers were published highlighting certain weaknesses of this scheme [28] [26]. Similarly, other protocols were presented later [11, 14, 29], but are also considered insecure according to many articles [12, 20, 23].

3 Our Contributions

- In the related works overviewed above, the protocols have a minimum of four and up to six exchanged messages. This type of protocols requires a number of operations on the reader side that is linear in the number of tags in the system, which might be impracticable since RFID systems can consist of a great number of tags. Our protocol needs only two message exchanges to fully complete the mutual authentication process.
- In all the protocols in Section 2 the tag keeps sending the same pseudo-dynamic index (IDS) as long as the reader has not been authenticated, which might make the protocol vulnerable to malicious traceability. In our scheme the tag does not send any message unless the reader is authenticated. Moreover all the exchange messages are randomized thanks to the nonce generated by the reader.
- It is not easy for low-cost RFID tags to install a PRNG, and hackers can also easily attack a system if PRNGs are not well designed. That is why we choose to implement the pseudo random number generator

(PRNG) on the server side to enhance security, and also to reduce the computation and storage requirements on the tag side.

- Unlike previous protocols in the state-of-art our protocol is not vulnerable to desynchronization attacks, as no updates are needed. Instead we use the nonce generated by the reader as a one-time-use authentication key.
- The proposed protocol relies on simple bitwise operations and does not involve computationally expensive cryptographic operations to provide protection against several attacks.

4 Description of the Proposed Scheme

4.1 Preliminaries and Notations

Three entities are involved in this scheme: the tag, the reader, and the backend server. Each tag stores the identifier IDT and the timestamp value t_i of the last authentication session to prevent replay attacks. Whereas the pseudo random number generator (PRNG) is only installed on the sever, therefore the tag needs only simple bitwise operations like XOR ($\oplus$) and left rotation ($Rot(A, B)$). Finally, the key K is shared between both tags and readers. We consider the length of the key and identifier is equal to 96-bits, which is compliant with EPCGlobal standard [1].We also assume that the communication between the reader and the back-end database is secure. On the contrary, the channel between the tag and the reader is considered insecure due to its open nature. The notations used throughout this paper are as follows:

- IDT: the identifier of the tag.
- K: a shared secret key
- R_i: a random integer
- $\oplus$: a bitwise XOR operation
- $Rot(A, B)$: a circular shift on A, ($B \bmod N$) positions to the left for a given value of N (in our case 96).
- t_i: the timestamp of the current session, typically giving date and time of day, this sequence of encoded information is used in order to identify when the communication has been initiated. This data is usually obtainable in a consistent format, allowing for easy comparison of two different records and tracking progress over time.

For the sake of clarity, we denote both the server and the reader as reader. The messages exchanged in the protocol are described below:

Step 1 *The reader sends a request message to the tag.*
The reader generates a random integer R_i, then computes A_i, B_i and C_i by

$$A_i = Rot(Rot(R_i, K) \oplus K, K) \tag{1}$$

$$B_i = Rot(Rot(K \oplus R_i, R_i), R_i) \tag{2}$$

$$C_i = Rot(Rot(t_i, K) \oplus R_i, R_i) \tag{3}$$

$A_i||B_i||C_i$ are sent along with the request message to the tag.

Step 2 *The tag authenticates the reader.*
Upon receiving $A_i||B_i||C_i$ as a query, the tag extracts R_i of A_i by

$$R_i = Rot^{-1}(Rot^{-1}(A_i, K') \oplus K', K') \tag{4}$$

Next, the tag verifies if K equals K' stored in its memory using R_i and B_i as follows:

$$K = Rot^{-1}(Rot^{-1}(B_i, R_i), R_i) \oplus R_i \tag{5}$$

Finally, the tag obtains t_i from C_i, and verifies if t_i is superior to t_i' stored during the last communication. The reader is authenticated if the message $A_i||B_i||C_i$ contains the right key K and the right timestamp t_i. The tag replaces then t_i' by the new t_i and computes $D_i||E_i$ to send it as a response by

$$D_i = Rot(Rot(K \oplus R_i, R_i), R_i) \tag{6}$$

$$E_i = Rot(Rot(IDT \oplus R_i, R_i), R_i) \tag{7}$$

Step 3 *The reader authenticates the tag.*
After receiving the response $D_i||E_i$, the reader computes D_i' by

$$D_i' = Rot(Rot(K \oplus R_i, R_i), R_i) \tag{8}$$

The reader will authenticate the tag if $D_i' = D_i$. Next the IDT is retrieved using

$$IDT = Rot^{-1}(Rot^{-1}(E_i, R_i), R_i) \oplus R_i \tag{9}$$

At this point, both the reader and the tag are mutually authenticated. The proposed ultralightweight RFID authentication protocol can be easily understood by looking at Figure 1.

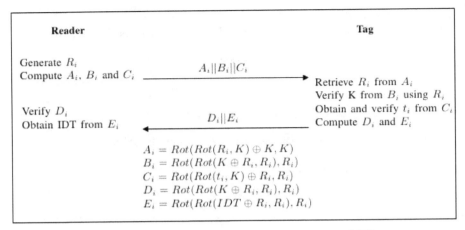

Figure 1   An ultralightweight security protocol for low-cost RFID tags.

5 Security Analysis and Performance Evaluation

The proposed protocol provides mutual authentication, the user's privacy protection and the resistance against various attacks such as impersonation and replay attacks. Moreover, the computational cost is quite low since only bitwise and rotation operations are used. A comparison of the relevant protocols in Section 1 is listed in Table 1.

5.1 Security Analysis

5.1.1 User's Privacy Protection

Traceability is one of the most difficult problems to solve. Indeed, malicious traceability allows recognizing and tracing an object or a person in different times and places. For instance, an attacker could identify important user's personal belongings in order to steal them, or track an important political person etc. Unlike the previous proposed protocols presented in Section 1, the tag does not send any message in our proposal, unless the reader is authenticated. Moreover, the double rotation provides a solid wrapping for the identity of the tag. In addition, all forward messages contain a random number which is different for each session. Therefore, it is difficult for an adversary to identify or trace a tag.

Table 1 Comparison of ultralightweight authentication protocols.

Protocols	UMAP Family [18, 19, 22]	SASI [7]	Gossamer [21]	This work
Total message exchanges	4	4	4	2
Resist de-synchronization attacks	×	×	×	✓
Resist disclosure attacks	×	×	✓	✓
Required memory on the tag[1]	6L	7L	7L	3L
User's privacy protection	×	×	✓	✓
Forward secrecy	×	×	✓	✓
Mutual authentication	×	✓	✓	✓
Operations in the tag	$\wedge, \vee, \oplus, +$	$\wedge, \vee, \oplus, +, Rot$	$\wedge, \vee, \oplus, +, Rot, MixBits^2$	$\oplus, Rot$

[1]L denotes the bit length of the secret information.

[2]MixBits is a lightweight pseudo random number generator.

5.1.2 Mutual Authentication

This feature is important for many applications, such as e-bank payments. Indeed, the proposed protocol provides mutual authentication; thus, only a legitimate reader possessing the key K can build a valid message $A_i||B_i||C_i$. Similarly, only a genuine tag can derive the random number R_i from $A_i||B_i||C_i$, and then compute message $D_i||E_i$. Also the exchanged messages involve secret values K and R_i that allow data integrity to be checked.

5.1.3 Forward Secrecy

Forward security feature guarantees the security of past communications, even if the tag is compromised later. In fact, even if current messages are exposed, the random numbers used to create all the exchanged information are different and unknown, which prevents from inferring any secrets from previous sessions. Thus, this protocol obtains forward secrecy, and reduces the chances of using previous sessions to compromise the communication between the reader and the tag.

5.1.4 Resist Impersonation Attack

For example, an attacker can try to be authenticated as someone else, and gain access to restricted areas without being authorized to do so. Also, an expensive object can be disguised into a cheap one. Thus, in the proposed scheme, if an adversary wants to deceive a tag, and pretends to be a legal reader, the attack would not be successful, because the secret key K is unknown and therefore A_i, B_i and C_i cannot be found. In the same way, an attacker would fail to deceive the reader since both R_i and K are unknown; thus, D_i and E_i cannot be forged to pass the authentication.

5.1.5 Resist Peplay Attack

The proposed protocol is designed to counter replay attacks. To our best knowledge, this is the first time that a timestamp information is included in the request messages to prevent this type of attacks in low-cost RFIDs. For example, an eavesdropper could try to impersonate the reader and replay the request $A_i||B_i||C_i$; however, the message would not be validated as it will not pass the verification of K and t_i. Similarly, if an adversary tries to replay the response $D_i||E_i$, the reader will detect the attack, because different random numbers are used in each session. Therefore, our protocol resists replay attacks.

5.2 Performance Evaluation

5.2.1 Low Computation Cost

Low-cost RFID tags cannot afford to use classic cryptographic primitives, mainly because hash functions and standard cryptographic algorithms have a very high computational cost and need large memory space. Consequently, these methods are not suitable for very constrained devices like low-cost RFIDs.

The protocol we have proposed only requires left rotation and simple bitwise XOR. The computational cost of these operations is quite low and can be implemented in hardware efficiently.

5.2.2 Communication Cost

All the protocols described in Section 2 have a minimum of four and up to six exchanged messages. This type of protocols may be impractical, since it requires a number of operations on the reader side that is linear in the number of tags in the system, and might be considered unfeasible since RFID systems are usually composed of a great number of tags. However, in the proposed protocol, the mutual authentication and the integrity protection are fully realized with only two messages.

5.2.3 Storage Requirement

Each tag needs to store only two records in ROM. The identity IDT and the key K are considered to have 96-bit length, which is compliant with all encoding schemes (i.e. GID, SGTIN, SSCC) defined by EPCGlobal standard [1]. The timestamp t_i is stored in a rewritable memory because it needs updates. Again, this is considered very low compared with the previous protocols described in Section 1.

6 Conclusion

The Radio Frequency Identification technology is nowadays used in many sensitive areas, but due to market considerations, the price has to be kept low. However, low-cost RFIDs are very constrained devices, and cannot support classic cryptographic primitives.

In this paper, we presented an ultralightweight protocol suitable for low-cost RFID tags. Furthermore, the proposed protocol resists several attacks such as malicious traceability and replay attacks. Also, the protocol needs only two forward messages to fully complete the authentication and identification process. Indeed, it is highly unlikely that the communication would be compromised, even though the key is shared. In fact, the double rotation offers a solid wrapping for the key; furthermore, all the messages are completely different thanks to the random number. As a result, the forward messages are not fixed; so an attacker cannot trace or identify a tag. Finally, the timestamp is used to provide a double security check and protects the tag from replay attacks.

References

[1] EPC Global Standard Class-1 Generation 2. http://www.epcglobalinc.org/standards/.

[2] B. Alomair, L. Lazos, and R. Poovendran. Passive attacks on a class of authentication protocols for rfid. In Proceedings of the 10th International Conference on Information Security and Cryptology, pages 102–115. Springer-Verlag, 2007.

[3] G. Avoine, X. Carpent, and B. Martin. Privacy-friendly synchronized ultralightweight authentication protocols in the storm. Journal of Network and Computer Applications, 2011.

[4] M. Bárász, B. Boros, P. Ligeti, K. Lója, and D. Nagy. Breaking LMAP. Proc. of RFIDSec, 7:11–16, 2007.

[5] M. Bárász, B. Boros, P. Ligeti, K. Lója, and D. Nagy. Passive attack against the M2AP mutual authentication protocol for RFID tags. In Proceedings of First International EURASIP Workshop on RFID Technology, 2007.

[6] T. Cao, E. Bertino, and H. Lei. Security analysis of the sasi protocol. IEEE Transactions on Dependable and Secure Computing, 6(1):73–77, 2009.

[7] H. Y. Chien. SASI: A new ultralightweight rfid authentication protocol providing strong authentication and strong integrity. IEEE Transactions on Dependable and Secure Computing, 4(4):337–340, 2007.

[8] H. Y. Chien and C. W. Huang. Security of ultra-lightweight rfid authentication protocols and its improvements. ACM SIGOPS Operating Systems Review, 41(4):83–86, 2007.

[9] P. D'Arco and A. De Santis. From weaknesses to secret disclosure in a recent ultralightweight rfid authentication protocol. Technical Report, Cryptology ePrint Archive. http://eprint.iacr.org/2008/470, 2008.

[10] P. D'Arco and A. De Santis. On ultralightweight RFID authentication protocols. IEEE Transactions on Dependable and Secure Computing, 8(4):548–563, 2011.

[11] M. David and N. R. Prasad. Providing strong security and high privacy in low-cost RFID networks. In Security and Privacy in Mobile Information and Communication Systems, pages 172–179, 2009.

[12] J. Hernandez-Castro, P. Peris-Lopez, R. Phan, and J. Tapiador. Cryptanalysis of the David–Prasad RFID ultralightweight authentication protocol. In Radio Frequency Identification: Security and Privacy Issues, pages 22–34, 2010.

[13] J. C. Hernandez-Castro, J. M. E. Tapiador, P. Peris-Lopez, and J. J. Quisquater. Cryptanalysis of the sasi ultralightweight RFID authentication protocol with modular rotations. In ArXiv e-prints, 2008.

[14] Y. C. Lee, Y. C. Hsieh, P. S. You, and T. C. Chen. A new ultralightweight RFID protocol with mutual authentication. In Proceedings of WASE International Conference on Information Engineering (ICIE'09), volume 2, pages 58–61. IEEE, 2009.

[15] T. Li and R. Deng. Vulnerability analysis of EMAP – An efficient RFID mutual authentication protocol. In Proceedings of the Second International Conference on Availability, Reliability and Security (ARES2007), pages 238–245. IEEE, 2007.

[16] T. Li and G. Wang. Security analysis of two ultra-lightweight RFID authentication protocols. In New Approaches for Security, Privacy and Trust in Complex Environments, pages 109–120, 2007.

[17] T. Li, G. Wang, and R. H. Deng. Security analysis on a family of ultra-lightweight RFID authentication protocols. Journal of Software, 3(3):1–10, 2008.

[18] P. Peris-Lopez, J. Hernandez-Castro, J. Estevez-Tapiador, and A. Ribagorda. Emap: An efficient mutual-authentication protocol for low-cost RFID tags. In Proceedings of OTM 2006 Workshops on the Move to Meaningful Internet Systems, pages 352–361. Springer, 2006.

[19] P. Peris-Lopez, J. Hernandez-Castro, J. Estevez-Tapiador, and A. Ribagorda. M2AP: A minimalist mutual-authentication protocol for low-cost RFID tags. In Ubiquitous Intelligence and Computing, pages 912–923, 2006.

[20] P. Peris-Lopez, J. Hernandez-Castro, R. Phan, J. Tapiador, and T. Li. Quasi-linear cryptanalysis of a secure RFID ultralightweight authentication protocol. In Information Security and Cryptology, pages 427–442. Springer, 2011.

[21] P. Peris-Lopez, J. Hernandez-Castro, J. Tapiador, and A. Ribagorda. Advances in ultralightweight cryptography for low-cost RFID tags: Gossamer protocol. In Information Security Applications, pages 56–68, 2009.

[22] P. Peris-Lopez, J. C. Hernandez-Castro, J. M. Estévez-Tapiador, and A. Ribagorda. LMAP: A real lightweight mutual authentication protocol for low-cost rfid tags. In Proceedings of 2nd Workshop on RFID Security, page 6, 2006.

[23] P. Peris-Lopez, J. C. Hernandez-Castro, J. M. E. Tapiador, and J. C. A. van der Lubbe. Security flaws in a recent ultralightweight RFID protocol. Arxiv preprint arXiv:0910.2115, 2009.

[24] C.M. Roberts. Radio frequency identification (RFID). Computers & Security, 25(1):18–26, 2006.

[25] H. M. Sun, W. C. Ting, and K. H. Wang. On the security of Chien's ultralightweight RFID authentication protocol. IEEE Transactions on Dependable and Secure Computing, 8(2):315–317, 2011.

[26] D. Tagra, M. Rahman, and S. Sampalli. Technique for preventing dos attacks on RFID systems. In Proceedings of International Conference on Software, Telecommunications and Computer Networks (SoftCOM), pages 6–10. IEEE, 2010.

[27] R. Weinstein. RFID: A technical overview and its application to the enterprise. IT professional, 7(3):27–33, 2005.

[28] K. H. Yeh and N. W. Lo. Improvement of two lightweight rfid authentication protocols. Information Assurance and Security Letters, 1:6–11, 2010.

[29] K. H. Yeh, N. W. Lo, and E. Winata. An efficient ultralightweight authentication protocol for RFID systems. Proceedings of RFIDSec Asia, 10:49–60, 2010.

Biographies

Wissam Razouk received her B.Sc and M.Sc degrees from Hassan II Universiy, Casablanca, Morocco. She is currently a PhD student in the Mathematics and Computer Science Department, in the same university. Her primary research interests are RFID security protocols and formal verifications.

Abderrahim Sekkaki received his D.Sc. in the Network Management domain from the "Paul Sabatier" University, France, 1991. He received a Dr. of State Degree from Hassan II University, Morocco in 2002 and is presently a computer science professor at the same university. His research interests include distributed systems, policies based network management, and security.

Efficient Discovery and Recovery of Common Control Channel in Cognitive Radio Wireless Ad-hoc Networks

Munam Ali Shah, Sijing Zhang and Carsten Maple

Department of Computer Science & Technology, University of Bedfordshire, Luton, LU1 3JU, UK; e-mail: {munam.shah, sijing.shang, carsten.maple}@beds.ac.uk

Received 8 February 2013; Accepted 13 February 2013

Abstract

The most important aspect of cognitive radio (CR) networks is to search, scan, and access the control channel to advertise the free channel list (FCL) amongst the participating CR nodes. Subsequent communication could not take place until there is an access to a well-known and agreed upon control channel to dialogue the initial configuration. In this paper, a novel protocol for searching, scanning, and accessing the control channel is proposed. The protocol consists of two levels of selection: rapid channel accessing and reliable channel accessing. In rapid channel accessing, nodes quickly and efficiently converge to a newly found control channel. In reliable channel accessing, switching to the backup control channel is performed when necessary. Furthermore, our reliable channel accessing allows CR nodes to access more than one control channel simultaneously. We evaluate the performance of the proposed approach through analytical modelling. The performance results show that our protocol can achieve efficient channel access time and fairness.

Keywords: MAC protocols, common control channel, co-operative communication, channel searching.

Journal of Green Engineering, Vol. 3, 273–289.

1 Introduction

The most important resource for wireless communications is radio spectrum [1]. Recent studies have revealed that spectrum is largely under-utilized most of the time [2]. Cognitive radio (CR) technology [3–5] has emerged as a promising solution to address the problem of spectrum shortage and inefficiency of its utilization in wireless networks. The CR technology significantly increases the spectrum efficiency by giving opportunistic access of the frequency bands to the unlicensed users (also called CR users or secondary users, SUs). Licensed users (also called primary users, PUs), while not transmitting, create spectrum holes, or white spaces in the spectrum. SUs use these white spaces owned by the PUs and opportunistically access these spectrum holes without interfering to PUs. In this paper, a decentralized scenario in ad-hoc cognitive radio networks (CRN) is considered.

A control channel is required by CR nodes to exchange the free channel list (FCL) and to dialogue initial configuration [6, 15]. Before SUs could start sending and receiving data, they exchange information on the control channel. This information could include sending and receiving FCL requests, availability of a spectrum hole and the time to be taken for the communication to last. CR nodes must have the capability to identify the characteristics of an unoccupied channel such as its available time and bandwidth, etc. Since PUs could arrive on their own spectrum bands, SUs must be able to sense the PUs claim in time and must quit the transmission on the occupied channels to avoid interference to PUs. If a PU claim is sensed during a communication, CR nodes must suspend their transmission. Consequently, CR nodes have to switch to another unoccupied channel to resume the transmission or re-dialogue configuration on control channel to agree upon a new white space for subsequent communication. Therefore, channel sensing and channel accessing are two main operations of CRN. Channel sensing deals with information about vacant channels in the environment, creating FCL and detecting PU presence while spectrum accessing is to exchange control information on a well-known control channel and to transmit data on a white space before a PU claims. Channel sensing is the task of physical layer and would be beyond the scope of this paper. For the rest of this section, we discuss the design issues of channel sensing and channel accessing and then review some of the related studies.

1.1 Design Constraints

Design constraints for channel accessing for CR users include efficiency of control channel, efficiency of data channel and efficiency of vacating a channel.

- Efficiency of control channel: This is reflected by the time required for CR nodes to discover a common control channel. Subsequent communication amongst CR nodes could not occur until CR nodes are aware of a control channel that is available for all CR nodes. The control channel efficiency depends on the selection criteria for control channel. The control channel could be either well known and publically available, commonly called global common control channel (GCCC) which is usually in industrial scientific and medical (ISM) band, or it could be one of the most reliable and available white spaces (non-GCCC). The former category is 24/7 freely available for any wireless application with no licensing issues but suffers from the drawbacks such as saturation of the GCCC (since it is widely available for anyone, which imposes high computational cost from backing off), no traffic differentiation (QoS unaware) and security attacks like denial-of-service (DoS). The latter category of control channel has worse searching efficiency, but once the control channel is discovered by all CR nodes in the vicinity, nodes spend less time in exchanging control information and get ready quickly to transmit data. Some of the researchers do not delve into selection criteria of control channel and simply assume that a control channel has already been found and established [7, 8]. This assumption is too strong because finding a control channel is a challenging task in CRN and actual data transmission could only take place until a successful and secure FCL transaction has taken place on a well-known and agreed-upon control channel.
- Efficiency of data channel: Data channel efficiency is defined as the time required for two CR nodes to conclude transmission on a data channel. In high traffic loads of PUs, CR users send only one data frame and then vacate the channel. However, when the chances of PU claiming are low and CR nodes still have data to send, more than one data frames will be transmitted in one transaction. The data channel efficiency could be increased by using more than one data channel simultaneously [9, 10]. On the other hand, determining the length of a spectrum hole could also help increase data channel efficiency.

- Efficiency of vacating a channel: CR users must vacate the occupied channel when the PU claims in order to minimize the interference. The majority of the CR MAC protocols found in literature assume that nodes are aware of the presence of PUs. However, the unrealistic assumption is criticized because CR nodes cannot sense PU presence when transmitting and PUs cannot generate interruptive signals to SUs on occupied channels. The performance of both PUs and SUs largely depends on whether or not the PU activity can be sensed in a timely manner. Equipping CR nodes with sensors in conjunction with transceivers could help alleviate the assumption and are less costly than transceivers [7].

2 Related Work

One of the challenging issues in CRN is to design an efficient MAC protocol that is capable of empowering the cognitive radio systems to handle changes at physical layer, eliminating the collisions as much as possible to avoid frame retransmissions, saving mobile energy and improving the network throughput by routing the packets to the destination with the minimal delay. Since the inception of CRN, a number of MAC protocols have been designed and developed. CR MAC protocols make use of either GCCC or non-GCCC to exchange control information before they can actually start communication.

Cognitive radio-enabled multi-channel MAC (CREAM-MAC) [7] is a decentralized CR MAC protocol that applies a four-way handshake with communicating nodes on the control channel under the assumption that the control channel is always available and reliable. Emphasis has been given on data transmission with complete ignorance of the overheads of determining and agreeing upon the control channel. It is strongly believed that finding a common channel to dialogue on the exchanged control information is the primary task of cognitive nodes, and that subsequent operations could not take place if the existence of the control channel has not been addressed. So the assumption of control channel being always available is not a well-built justification.

In opportunistic-cognitive MAC (OC-MAC) [8], initially all nodes reside on a non-global common control channel, perform three-way handshakes to select a data channel from the FCL, and then confirm the data transmission through an acknowledgement. CR nodes in OC-MAC predict the length of spectrum hole, but this prediction is strongly criticized because the CR network is an opportunistic network and it is very hard to find the exact length of time during which the PU is not utilizing the spectrum so that the length of the available spectrum hole could be calculated.

The cognitive MAC protocol using Statistical Channel Allocation for wireless ad-hoc networks (SCA-MAC) [9] is a decentralized GCCC-based CR MAC protocol that can speed up transmission by using more than one channel for data transmission and can wait for some time for a channel with higher bandwidth to become available. Again, the protocol emphasizes on data transmission and ignores the pre-transmission overheads such as the time required in dialogue to exchange initial configuration and the time required to converge on the common control channel.

Cognitive MAC (C-MAC) protocol for multi-channel wireless networks [11] selects the so-called R channel within the white spaces and sets this channel as a control channel and manages the communication on R channel. However the selection criterion for the R channel has not been clearly defined, and also the clarification about which node will select the R channel and how the rest of nodes will be synchronized is missing.

An efficient MAC protocol for improving the throughput for CR networks (A-MAC) [10] belongs to a decentralized non-GCCC family of CR MAC protocols, in which the spectrum sensing is done using a half-duplex transceiver before the channel state information is made available to nodes. The protocol exchanges the FCL with communicating partners on the most reliable control channel. Although the results provided show that a higher throughput is achieved, a clear description of the mechanism used by CR nodes to converge on the reliable channel is missing. Also, a clear methodology is required to address the hidden terminal problem using a half-duplex transceiver.

Dynamic Open Spectrum Sharing MAC (DOSS-MAC) protocol for wireless ad-hoc networks [12], which makes use of three transmitters, presents a control channel algorithm to enable coordination among cognitive nodes and implements network layer multicasts. The hidden terminal problem is efficiently addressed in DOSS-MAC using three transceivers. It is suggested that similar functionality could be achieved using sensors instead of transceivers which could be a more cost effective solution.

SYNChronized MAC (SYNC-MAC) protocol for multi-hop cognitive radio networks [13] chooses one of the channels common between itself and neighbours to exchange control signals while other channels are selected to send data. It is not possible to decide a common channel until a CR pair have exchanged their FCLs.

2.1 Objectives

This paper aims to design a dynamic, decentralized, and hybrid medium access control protocol, named DDH-MAC for an overlay ad-hoc CRN. The protocol is dynamic because whenever a PU claim happens, CR nodes efficiently agree upon a newly found control channel to maintain control channel efficiency. The architecture of the protocol is decentralized, not infrastructure-based. DDH-MAC is hybrid in nature making partial use of both GCCC and non-GCCC families of CR MAC protocols. The framework for DDH-MAC has been presented in our previous work [14]. In this paper, we enhance our research by introducing the multi-layer reliability factor and we present an efficient and robust control channel access mechanism that emphasizes on the control channel efficiency. CR nodes implementing the proposed mechanism are always in a state where they have access to at least one control channel even after the PU interference has been sensed. CR users in the proposed mechanism, without renegotiations, switch to another control channel whenever there is a PU claim. CR nodes have access to three control channels at the same time. This unique feature smartly and intelligently addresses the PUs' channel re-occupancy, reduces the impact of re-exchange of control information and leads towards reliable communication in CRN.

The rest of the paper is organized as follows. The detailed operation of proposed scheme has been described in Section 3. The reliability factor and efficiency of the protocol have been computed through some mathematical calculations in Section 4. Section 5 discusses results before the paper is concluded in Section 6.

3 An Efficient and Robust Decentralized Control Channel Access Mechanism for CRN

In this section, we make some assumptions, define control frames, and then describe DDH-MAC phases and the operation. The protocol sets the following assumptions.

- Each CR node is equipped with two transceivers: G-Transceiver (GT) to continuously and rapidly scan global control channel, and D-Transceiver (DT) to transmit data.
- CR nodes utilize the CSMA/CA mechanism to access control channel.
- Spectrum has been sensed by the physical layer and FCL has been populated by each CR node.

- Each CR node is equipped with two sensors: one sensor senses the PU activity on a local control channel (PCCH or BCCH) and other sensor senses the PU activity on the data channel.

Four control messages are exchanged in DDH-MAC. One control message is delivered through GCCC.

1. Beacon Frame (BF) is launched in GCCC by the first node in the CRN to inform all the other CR nodes about the primary control channel (PCCH) and backup control channel (BCCH). Both PCCH and BCCH are one of the white spaces and will be used as a local control channel (non-GCCC). Two parameters are carried in BF: channel ID of PCCH and channel ID of BCCH. Channel IDs are arbitrarily selected by the first CR node, and are the first two channels that the CR node has sensed. Channel ID has a numeric value, where $0 \leq$ Channel ID $< N$.
   The local control channel (PCCH or BCCH) delivers three types of control frames.
2. DDH-MAC Control Frame (DMCF) is utilized by a potential CR sender to inform all the CR nodes in the vicinity that it is ready for communication.
3. Free channel list (FCL) is utilized by the same CR sender who sent DMCF which includes channel IDs of all channel that could possibly be used as data channels for subsequent transmission.
4. ACK is utilized by a CR receiver, who wins the contention on PCCH. The receiver replies with its own FCL identifying the channels common between CR pair for possible data transmission.

DDH-MAC consists of two phases: rapid channel accessing and reliable channel accessing.

3.1 Phase 1: Rapid Channel Accessing

When a CR node wants to transmit data, it first scans the GCCC for BF. There are two possibilities:

1. If any BF is found (Figure 1②), the information about PCCH and BCCH is learnt. This also means that the node will join an existing CRN and now PCCH needs to be scanned to learn more about the network.
2. If the CR sender does not find any BF in GCCC, then this node becomes the first CR node in CRN and is responsible for three functions: setting one of the white spaces in its FCL as PCCH and another as BCCH; form-

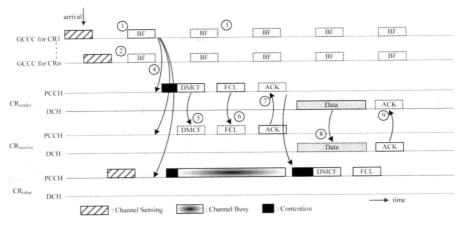

Figure 1   An example of Phase 1 operations.

ing and launching BF in GCCC (Figure 1①); and keeping transmitting copies of BF at regular intervals (Figure 1③).

In both cases, the CR node starts scanning the PCCH and observes the activities on the local control channel (Figure 1④). CR sender and receiver then exchange three control information frames through PCCH. Firstly, DMCF is launched (Figure 1⑤), followed by transmitting the FCL (Figure 1⑥). DMCF and ACK also serve to avoid the hidden terminal problem which is traditional in ad-hoc networks. The intended recipient checks its FCL to see if a common channel exists. If a common channel is found, a reply with an ACK is sent to the sender (Figure 1⑦). The pair then switch to the identified common data channel and start transmitting data using DT (Figure 1⑧). All the data frames are acknowledged using data ACK (Figure 1⑨). Other nodes will wait for PCCH to become idle and will contend to dialogue the control information after it is sensed free. GT will be used by all CR nodes in the network to scan the local control channel to have knowledge about all the activities carried out by other CR nodes in the network. The CR pair which just finished communication could remain unaware of the status of other CR nodes, and thus continuously scanning the control channel helps track the record of other CR nodes' activities. This ultimately avoids the hidden terminal problem. In rapid channel accessing, nodes can access the control channel efficiently and rapidly. Any new node joining the network firstly searches for a beacon frame which could be read for information about local control channel(s). After this, nodes simply switch to the newly discovered control channel for

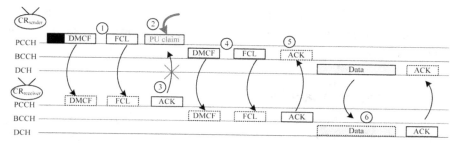

Figure 2 Phase 2 operations.

the most crucial part of communication i.e., FCL transactions, which leads to data frames transmission.

3.2 Phase 2: Reliable Channel Accessing

The PCCH and BCCH make use of the most readily available white spaces scanned and setup by a CR user. Unlike GCCC, which is publically available to everyone and is more prone to security vulnerabilities, FCL could be exchanged privately and secretly amongst CR nodes through PCCH after the nodes in the vicinity has converged on this local control channel. Since the PU can claim any occupied channel any moment of time, PCCH could also be claimed and as a result, nodes using PCCH for control information have to either abort the configuration dialogue or renegotiate on other white spaces. The proposed mechanism efficiently deals with this situation by using BCCH to resume the exchange of control information if there is a PU claim on PCCH. Figure 2 illustrates the example of Phase 2 where the CR sender is transmitting control information on PCCH (see Figure 2①) and is awaiting ACK from the recipient. Meanwhile, a PU claim is sensed on PCCH (see Figure 2②) due to which ACK could not be delivered (Figure 2③).

CR nodes can switch to BCCH without re-negotiations and re-searching control channel, and resume transceiving the control information (see Figure 2④⑤), followed by the data transmission on an agreed data channel (see Figure 2⑥). In the worst case scenario when BCCH is also claimed by the PU, CRN will execute operations of Phase 1 and will converge on new PCCH and BCCH. This dynamicity of local control channels provides the nodes an extra security feature. An adversary, planning to attack PCCH or BCCH and manipulating information on control channel, has to re-compile the attack every time when new PCCH and BCCH are set up. The reliable channel access gives CR nodes the assurance that they always have access to three channels simultaneously and any channel could be used for subsequent exchange of control information.

4 Performance Analysis

In this section, we first discuss different case scenarios, and then model the process of control channel efficiency and reliable channel efficiency, and finally calculate the time it takes for exchange control information.

As previously discussed, the proposed scheme performs a few operations before the network is fully converged. These operations include scanning/sensing GCCC, exchanging FCL on PCCH or BCCH (if there is a PU claim on PCCH) and lastly concluding transmission on the agreed white space(s). Each of the above listed operations requires time for its completion such as time required to sense/scan BF in GCCC, time required to launch BF, time to read BF and time required to exchange FCL on PCCH/BCCH. All these operations form part of pre-transmission time which heavily affect the throughput and QoS as nodes holding delay-sensitive data will be highly affected through varied values of pre-transmission time. Let T denote any of the above mentioned operations and T_{PT} represent the pre-transmission time which is further expressed as

$$T_{PT}^{DDH-MAC} = \{T_{BS}, T_{BF}, T_3, T_{FCL}^{PCCH}, T_{FCL}^{BCCH}, T_{DMCF}, T_{ACK}\} \quad (1)$$

where T_{BS} is the time required to scan GCCC for BF, T_{BF} is the time to read BF or launch BF in GCCC, T_3 is the waiting time before a CR node can launch BF. Note that this waiting also aims to avoid duplication of BF by multiple CR nodes. T_{FCL}^{PCCH} and T_{FCL}^{BCCH} are the amount of time a CR node takes to broadcast its FCL in PCCH or BCCH if there is a PU claim. T_{DMCF} and T_{ACK} are control frames similar to RTS/CTS and are used to avoid the traditional hidden terminal problem. They are exchanged between communicating nodes before actual transmission can take place and lastly, $T_{PT}^{DDH-MAC}$ denotes the Pre-Transmission time.

4.1 Case Scenarios in DDH-MAC

Not all the operations are performed by cognitive nodes in DDH-MAC, and the number of operations performed depends on the role of a CR node and the case scenario. Currently, there are four cases in DDH-MAC. Case I represents network initialization phase where no control channels have been found and

Table 1 The parameters for the proposed scheme.

Parameter	Assigned Value
BF	14 Byte
DMCF	20 Byte
FCL	20 Byte
ACK	14 Byte
T_{BS}	10.181 μs
T_{BF}	10.181 μs
TDMCF	14.545 μs
T_{FCL}^{PCCH}	14.545 μs
T_{FCL}^{BCCH}	14.545 μs
T_3	30.543 μs
T_{ACK}	10.181 μs

CR Node 1 creates and launches the BF in GCCC. Case II represents the scenario where CR nodes after scanning GCCC find BF, read information about local control channel and then switch to PCCH. In Case III, the network initialization phase in addition to PU claim on the PCCH is considered. The last case is Case IV which is extension of Case II, in which nodes, after scanning GCCC and finding information about the PCCH, are forced to switch to BCCH due to PU arrival on the PCCH. Based on the number of operations performed in each case scenario, T_{PT} for all the four cases has been derived, as shown in the following equations:

$$T_{PT1}^{DDH-MAC} = \{T_{BS} \cup T_3 \cup T_{BF} \cup T_{DMCF} \cup T_{FCL}^{PCCH} \cup T_{ACK}\} \qquad (2)$$

$$T_{PT2}^{DDH-MAC} = \{T_{BF} \cup T_{DMCF} \cup T_{FCL}^{PCCH} \cup T_{ACK}\} \qquad (3)$$

$$T_{PT3}^{DDH-MAC} = \{T_{BS} \cup T_3 \cup T_{BF} \cup T_{DMCF} \cup T_{FCL}^{PCCH}$$
$$\cup T_{ACK} \cup T_{DMCF} \cup T_{FCL}^{BCCH} \cup T_{ACK}\} \qquad (4)$$

$$T_{PT4}^{DDH-MAC} = \{T_{BF} \cup T_{DMCF} \cup T_{FCL}^{PCCH} \cup T_{ACK} \cup T_{DMCF} \cup T_{FCL}^{BCCH} \cup T_{ACK}\}$$
$$\qquad (5)$$

The above equations are used to compute the T_{PT} for DDH-MAC. We have used IEEE 802.11b as a benchmark to calculate values for above operations. Table 1 summarizes the values.

For simplicity, we have considered a static size for contention window, and channel conditions have been set to ideal. A total of 124 Bytes, 68 Bytes, 178 Bytes and 122 Bytes are exchanged in Cases I to IV of proposed scheme

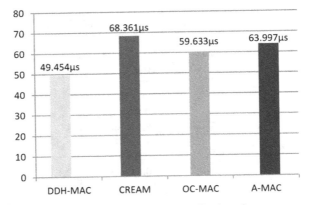

Figure 3   Case I: Network initialization phase.

respectively which yield following values of T_{PT}.

$$T_{PT1}^{DDH-MAC} = 90.178 \ \mu s$$

$$T_{PT2}^{DDH-MAC} = 49.454 \ \mu s$$

$$T_{PT3}^{DDH-MAC} = 129.447 \ \mu s$$

$$T_{PT4}^{DDH-MAC} = 88.727 \ \mu s$$

5 Results and Discussion

In this section, T_{PT} has been calculated for other CR MAC protocols [6–8] for performance comparison and evaluation. Figure 3 shows the T_{PT} for Case I. The obvious reason for the high value of T_{PT} is the fact that the network is in the initialization phase and Node 1 has to wait for a certain amount of time to avoid BF duplication. Since other protocols do not wait to launch BF and the network is initialized through scan activity (or under the assumption of the existence of available control channel) followed by exchange of control frames, T_{PT} is less for other protocols in Case I.

Case II in our protocol has the lowest T_{PT} of 49.454 μs when compared with other protocols. Nodes in DDH-MAC read the BF, simply switch to PCCH and contend to dialogue the control information (Figure 4).

The reliability of our scheme is revealed in Cases III and IV. The PU claim on control channel is efficiently addressed by switching to BCCH and resuming the exchange of control information. Unlike other protocols, CR

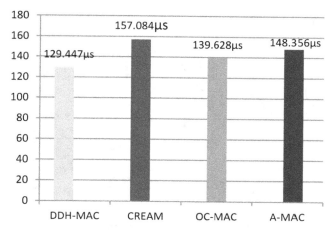

Figure 4 Case II: Network is initialized and control channel is found.

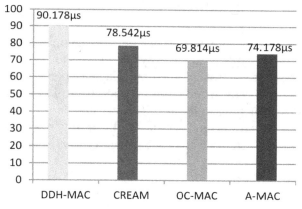

Figure 5 Case III: Network initialization phase, when PU interference is sensed in local control channel.

nodes in our protocol do not need to re-dialogue the entire configuration whenever the PU occupancy is detected. Figure 5 shows the efficiency of our scheme in Case III and also reveals that the more number of control frames are exchanged in CREAM-MAC, OC-MAC and A-MAC, yielding to high values of T_{PT}.

The proposed scheme outperforms other MAC protocols in case IV and consumes the least time before the nodes finish exchanging control information and start transmitting data on a data channel.

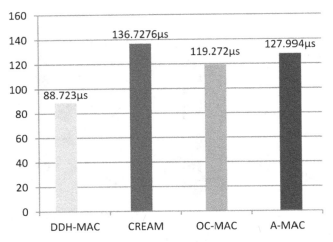

Figure 6 Case IV: The PU occupancy happens after the network is initialized, and control information is re-exchanged.

The T_{PT} for the scenarios of PU claiming is computed and expressed in Figures 5 and 6. The typical response by CR nodes is to abort the transmission and re-exchange the control information to agree upon another white space to conclude transmission while the proposed protocol efficiently deals with the situation by switching to the BCCH. Less number of frames exchanged with other CR nodes results in faster network convergence, and nodes remain in the state where at least one control channel remains always available to all CR nodes.

6 Conclusion

Cognitive radio networks aim to be a promising technology to resolve the problem of spectrum scarcity. CR nodes must exchange the control information on a control channel before transmitting the actual data. The selection criteria of CCC make the CR technology reliable. In this paper a novel multifold reliable framework for CR networks has been presented, which uses more than one control channel at the same time. Rapid channel access makes the CRN converged fast and rapid while reliable channel access gives the nodes assurance that they have access to at least one control channel to set up initial configuration dialogue. T_{PT} plays a very important role in the performance of the CRN. We have computed and compared the T_{PT} with some other

CR MAC protocols. The small values of T_{PT} lead to mobile energy efficiency as nodes have to wait less before the actual communication starts. Currently, the proposed mechanism is being simulated for evaluation and performance comparison of parameters like throughput, delay and energy consumption.

References

[1] The Radio Spectrum. Available online at: `http://transition.fcc.gov/Bureaus/OPP/working\_papers/oppwp38chart.pdf`.

[2] P. Kolodzy. Spectrum policy task force, Federal Communication. Commission, Washington, DC, Technical Report ET Docket, No. 02-135, 2002.

[3] J. Mitola and G. Q. Maguire. Cognitive radio: making software radios more personal. IEEE Personal Communications, 6(4):13–18, 1999.

[4] S. Haykin. Cognitive radio: Brain-empowered wireless communications. IEEE Journal on Selected Areas in Communications, 23(2):201–220, Feb. 2005.

[5] I. F. Akyildiz, W.-Y. Lee, M. C. Vuran, and S. Mohanty. Next generation/dynamic spectrum access/cognitive radio wireless networks: A survey, Computer Networks, 50(13):2127–2159, 2006.

[6] M. A. Shah, S. Zhang, and C. Maple. An analysis on decentralized adaptive MAC protocol for cognitive radio networks. In Proceedings 18th International Conference on Automation & Computing (ICAC'12), pp. 1–5, 2012.

[7] X. Zhang and H. Su. CREAM-MAC: Cognitive radio-enabled multi-channel MAC protocol over dynamic spectrum access networks. IEEE Journal of Selected Topics in Signal Processing, 5(1):110–123, Feb. 2011.

[8] S.-Y. Hung, Y.-C. Cheng, E. H.-K. Wu, and G.-H. Chen. An opportunistic cognitive MAC protocol for coexistence with WLAN. In Proceedings 2008 IEEE International Conference on Communications, pp. 4059–4063, 2008.

[9] A. C.-C. Hsu, D. S. L. Wei, and C.-C. J. Kuo. A cognitive MAC protocol using statistical channel allocation for wireless ad-hoc networks. In Proceedings 2007 IEEE Wireless Communications and Networking Conference, pp. 105–110, 2007.

[10] G. P. Joshi, S. W. Kim, and B.-S. Kim. An efficient MAC protocol for improving the network throughput for cognitive radio networks. In Proceedings 2009 Third International Conference on Next Generation Mobile Applications, Services and Technologies, pp. 271–275, 2009.

[11] C. Cordeiro and K. Challapali. C-MAC: A cognitive MAC protocol for multi-channel wireless networks. In Proceedings 2nd IEEE International Symposium on New Frontiers in Dynamic Spectrum Access Networks, pp. 147–157, 2007.

[12] L. Ma, X. Han, and C.-C. Shen. Dynamic open spectrum sharing MAC protocol for wireless ad hoc networks. In First IEEE International Symposium on New Frontiers in Dynamic Spectrum Access Networks (DySPAN 2005), pp. 203–213, 2005.

[13] Y. R. Kondareddy and P. Agrawal. Synchronized MAC protocol for multi-hop cognitive radio networks. In Proceedings 2008 IEEE International Conference on Communications, pp. 3198–3202, 2008.

[14] M. A. Shah, G. A. Safdar, and C. Maple. DDH-MAC: A novel dynamic de-centralized hybrid MAC protocol for cognitive radio networks. In Proceedings of RoEduNet, pp. 1–6, 2011.

[15] M. A. Shah, S. Zhang, and C. Maple. An analysis on decentralized adaptive MAC protocols for Cognitive Radio networks. International Journal on Automation and Computing (IJAC), 10(1):46–52, February 2013.

Biographies

Munam Ali Shah received his B.Sc and M.Sc degrees, both in Computer Science from University of Peshawar, Pakistan, in 2001 and 2003 respectively. He completed his MS degree in Security Technologies and Applications from University of Surrey, UK, in 2010, and is currently a final year PhD student at University of Bedfordshire, UK. Since July 2004, he has been a Lecturer, Department of Computer Science, COMSATS Institute of Information Technology, Islamabad, Pakistan. His research interests include MAC protocol design, QoS and security issues in wireless communication systems. Mr. Shah received the Best Paper Award of the International Conference on Automation and Computing in 2012.

Sijing Zhang obtained his B.Sc and M.Sc degrees, both in Computer Science, from Jilin University, Changchun, P.R. China in 1982 and 1988 respectively. He earned a Ph.D in Computer Science from the University of York, UK in 1996. He then joined the Network Technology Research Centre (NTRC) of Nanyang Technological University in Singapore as a Post-Doctoral Fellow. In 1998, he returned to the UK to work as a Research Fellow with the Centre for Communications Systems Research (CCSR) of the University of Cambridge. Dr. Zhang joined the School of Computing and Technology, University of Derby, UK, as a Senior Lecturer in 2000. Since October 2004, he has been working as a Senior Lecturer in the Department of Computer Science and Technology, the University of Bedfordshire, UK. His research interests include real-time scheduling algorithms, schedulability tests for hard real-time traffic, performance analysis and evaluation of real-time communication protocols, and wireless networks for real-time industrial applications.

Carsten Maple obtained his B. Sc. (Hons) degree at the University of Leicester, UK in 1994, and his Ph.D. degree in 1998. He joined the University of Luton, UK in 1998 as a lecturer in computer science before moving into senior lecturer role later the same year and a principal lecturer post by 2002. In 2004, he was appointed professor of applicable computing, and was promoted to head of the Department for Computer Science and Technology at the newly formed University of Bedfordshire, where he led the department from strength to strength before his promotion to pro-vice chancellor (Research and Enterprise) in 2010. His research interests include network and computer security, cyberstalking, ethical hacking, information security, trust and authentication in distributed systems, and optimization techniques.

Wireless Sensor Networks – Routing Impact on Energy Distribution and Energy Hole Formation

Damir Zrno[1], Dina Šimunić[1] and Ramjee Prasad[2]

[1]FER, University of Zagreb, Zagreb, Croatia; e-mail: damir.zrno@fer.hr
[2]CTIF, Aalborg University, Aalborg, Denmark

Received 29 March 2013; Accepted 2 April 2013

Abstract

In this paper we investigate the impact of different routing techniques on energy consumption in sensor networks with a central node. When multi-hop transmissions are used the sensors stationed closer to the central node will have to relay a large amount of traffic and expend their batteries faster. This leads to the "energy hole" problem ultimately resulting in reduced network life. We use a sensor network simulator to test the effectiveness of several routing techniques for homogenous networks and compare them to heterogenous networks using relay nodes to determine the best approach in terms of network life.

Keywords: Smart routing, sensor networks, energy hole problem, many-to-one communication model, energy consumption, green communication.

1 Introduction

A lot of research has been done recently on wireless sensor networks and this trend will continue as they are becoming a more and more important part of our life [1, 2]. Sensor networks are typically formed from a number of nodes in order to monitor a specific kind of activity. These range from home environment applications [3], mapping applications [4], weather and seismic measurements [1] or even battlefield monitoring. Such nodes are usually limited foremost by size and the need to operate without infrastructure like power lines or communication lines. This leads to fundamental limitations in battery power (network life), computational power and transmission capabilities (communication range). Much effort is directed today towards pushing these limitations [5–7], although battery power remains the biggest concern. This paper focuses on uneven energy consumption in large scale sensor many-to-one networks with a central data-gathering node.

Sensor networks can be categorized by their data traffic pattern into one-to-many or many-to-one networks. One-to-many network has each node disseminating its data towards other interested nodes. A special case is a many-to-many communication network where data is transferred between pairs of nodes. In a many-to-one sensor network all of the nodes are sending their sensing data towards a central node, typically a base station. This central node then either stores this information or carries specialized relay-communication equipment (such as a satellite dish or long range wireless modem). Such networks are typically used for environment monitoring and surveillance [8, 9].

Scalability problems in routing and energy consumption occur when such many-to-one networks are used to cover a large area requiring more than one hop in communication as well as a large number of devices in order to adequately monitor the environment. One of the problems occuring in such large scale sensor networks is an uneven energy consumption rate, particularly around the central node. Nodes close to the central one have to relay more traffic than the outlying nodes, resulting in faster energy consumption. These nodes are then the first to drain their energy supply leading to a dead battery area around the central node. This is called an "energy hole" problem [10].

Different routing techniques can partially mitigate this problem by sending some nodes to sleep in order to spread the routing load, however this does not completely remove it. Clustering and data aggregation have most often been proposed as a solution. Such networks can easily be scaled due to their multi-hierarchy architecture used to channel data to local relay nodes which then proceed to transmit information directly to the central base station or by relaying it between themselves [11, 12].

We have developed a sensor network simulator for real-case scenario simulations in both physical and mac layer, supporting many different routing techniques, including greedy forwarding, multi-hop smart routing, shotgun forwarding and flooding [13]. We use it to investigate the impact of the energy hole problem on different routing techniques and to investigate how clustering affects such networks. Of particular interest are data congestion and total network life in terms of time and reported events per unit of energy. We have done extensive simulations on the performance of different routing techniques, some of which are presented in [14]. This paper expands on that work and gives new simulations showing the effects of clustering.

Section 2 of this paper gives a short background on the energy hole problem and typical solutions used in sensor networks. Section 3 details simulation setup, assumptions and parameters used. Section 4 presents simulation results in terms of network lifetime, energy efficiency and battery usage through different parts of the network. Section 5 contains concluding remarks.

2 Background

2.1 The "Energy Hole" Problem

Data in many-to-one networks is transmitted to the central base station using multi-hop routing. If we assume maximum range R of an individual node, then nodes within R circle radius around the central base station can transmit their data with a single direct link (Figure 1). However, nodes with distance between R and $2R$ from the central node (second circle) have to transmit their information through the nodes within the first circle. This creates additional load on the first circle nodes as they need to spend additional energy to first receive the data packets and then to forward them towards the central node. It is clear that nodes within this circle will run out of battery power first, thus creating a no-pass zone around the central node and preventing any data gathering even if the central node is still operational and outlying nodes continue to monitor the environment. We assume here that packets can traverse the first circle using a single hop which may not always be the case in real-case scenarios.

Assuming same node density within both circles it follows that for each packet generated within the first circle there are three packets generated within the second circle as it has three times the surface area. That means that nodes within the first circle have to relay three times as many events as they themselves generate. This problem is further compounded if the nodes within first circle are not in sleep mode and have to receive parts of each

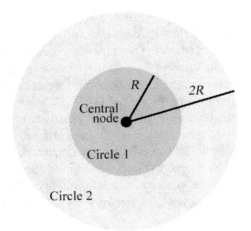

Figure 1 Simplified energy hole problem with nodes in first circle having to relay the second circle information on top of their own data.

incoming data packet before they can determine whether they belong to its multi-hop transmission path.

Sensor network lifetime is usually defined as the time period within which the network can maintain its functionality by covering the designated area and sending data with good reporting rate. In this paper it is defined as the time until the first node dies. It is of big importance to see where the first nodes cease to function and to compare that to the energy distribution through the network. A better network will drain the batteries of its nodes more evenly. Other big factors are to see how many events the network is able to report during its lifetime and how long can it keep actively monitoring the environment.

2.2 Energy Conservation Approaches

The very basic form of network uses unguided routing and simply spreads the sensed data to the whole network (flood routing). This is a very inefficient form of communication so other approaches like location aware routing where nodes forward the packet only if they are closer to the destination from the previous hop (location aware routing) are used, but this still results in a large number of excessive transmissions. Targeted transmissions such as greedy forwarding are typically employed in sensor networks. A node wishing to transmit uses maximum transmission power in order to send the

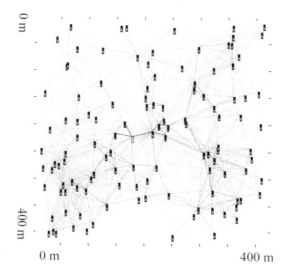

Figure 2 Smart routing in an ad-hoc network. Lines represent data transmission paths with stronger lines representing paths used more often.

packet as far towards its destination as it can. This however causes a lot of interference and often uses more power than necessary.

Smart routing algorithms consider network topology and select the best possible multi-hop transmission path, employing only the minimal required power in each of the hops. However, these are most computationally intensive out of the mentioned routing protocols and require dissemination of up-to-date network topology to operate efficiently. Figure 2 shows an implementation of smart routing in an ad-hoc network. Lines represent connections between the nodes, and strength (intensity) of each line shows how often it is used for communication – stronger lines are used more often. Lines typically run from outside nodes towards nearest nodes and the middle. Nodes are represented by small rectangle bars placed at their positions showing current battery status (the more empty the battery becomes, the lower its battery level indicator will be).

Deployment assistance in the form of clustering is often used to combat the energy hole problem. Several relay nodes are distributed around the network area, each receiving a sub- region from which to collect data and forward to the central node using its own (often enhanced) transmission power. These can be used in conjunction with any of the mentioned routing proto-

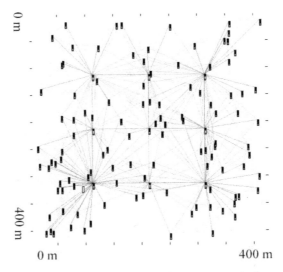

0 m

400 m

0 m   400 m

Figure 3 Routing in a hierarchial network. Connections run towards the nearest relay nodes from where they are routed towards the central node.

cols. In this paper we will focus on relay nodes using the same transceiver cirquits as the rest of the network, but having enhanced batteries to cope with the additional load. Such networks utilizing relay nodes are hierarchical in nature and have distinctive traffic patterns, as can be seen in Figure 3. Connections in such a network typically run from regular nodes towards the nearest relay node, and then from them towards the central base station.

3 EnVO Simulator and Network Parameters

EnVO network routing simulator was developed in MATLAB programming language with the purpose of testing the influence of various routing protocols in a realistic ad-hoc wireless network [15]. It was further developed into encompassing hierarchical approaches. EnVO routing simulator incorporates the common routing schemes – flood routing, shotgun routing, greedy forwarding, and two smart power control routing algorithms – one based on best hop routing, the other on two-hop path-finding. A general review on current routing methods in ad-hoc networks can be found in [13, 16]. It also enables simulation of hierarchical networks using local relay nodes.

Data transfer rate, environment interference levels, network throughput, node battery levels and network life are monitored using same network topography and simulation parameters with different routing protocols.

Several assumptions have been used in simulation:

- Nodes are randomly distributed over simulation area using a uniform distribution.
- Network is well connected.
- Routing tables have been established, relay nodes selected, and network data required for specific routing protocol disseminated to every node.
- Each sensor node starts in idle mode monitoring the environment and sensing radio traffic in its area.
- Events occur randomly through network with uniform distribution.
- Network is event driven, although simulation results are valid for time driven networks as events are generated at the same rate.
- Each sensor node has the same likelihood to notice an event while operating in monitoring mode. A report is generated in the form of a data packet and sent to the central node.
- All sensor nodes have the same transceiver circuits requiring same signal levels for successful reception.
- All regular nodes have the same batteries, fully charged, while relay nodes and the central node carry batteries with three times higher capacity.

In our simulations we consider a square area of 400 by 400 m with a base station located in the center of the map and 200 nodes spread through the environment equalling node density of 1 node per 800 m^2 (approximately 28 m average distance between neighbouring nodes). Same node locations have been used for all routing protocols.

Communication protocols include collision avoidance using radio sensing and random back-off time. Network operates using 1 Mbps data rate, with each event report carrying 1 kb of data. Events are randomly generated at the average rate of 10 events per second. Nodes are able to operate with transmission levels between 1 and 100 mW and have −95 dBW receiver sensitivity resulting in a maximum range of 169 m. Nodes are categorized into four concentric circles equalling half a hop distance radius increase (85 m) to better follow battery levels. The central node's battery levels are monitored separately.

Nodes have 1 J battery capacity, while central and other relay nodes have 3 J battery capacity. Energy consumption of each node is as follows:

- 1 mW while idle.
- 60 mW during packet negotiation sensing.
- 135 mW while receiving data.
- 45 mW for transmission processing.
- Between 1 and 100 mW during data transmission.

4 Simulation Results

Simulations have been run on the same network topology for each of the routing schemes to determine the best approach for large scale networks. Average battery levels relative to their starting value for each of the concentric circle areas around central node at the end of network lifetime are shown in Table 1. As expected, there is an uneven distribution of energy between circles, especially in the case of flood routing, shotgun routing and SLAM routing. Shotgun routing and SLAM routing in particular suffered network failure with still very high battery levels, indicating large variety in energy consumption between different nodes within the network. Both greedy forwarding and and optimal multi-hop routing benefitted from the use of clustering, spreading the load within the network more evenly.

Table 2 shows network parameters at the end of network lifetime. Time before first node battery failure measured in seconds is displayed in the first column. Greedy forwarding and optimal multi-hop routing are clearly the best routing methods regarding network lifetime, with similar results. Optimal multi-hop with clustering achieved the best network lifetime of 645 seconds. This is for a communication network operating in 5% load range with 1 J stored energy per battery at the start of simulation. The number of reported events during network life, both as absolute and as a percentage of actual occured events is given in column 2. It is clear from this that flood routing did not benefit from clustering, quite the opposite. Shotgun routing had a small reduction in required energy per reported event, but also performed badly with less than half events reported. These are the results of a too high network load for these routing protocols. It is clear that they cannot handle a 5% network load (average data stream of 50 kbps in a 1 Mbps network). Greedy forwarding and optimal multi-hop performed much better and benefited further from clustering. Optimal multi-hop in particular achieved a very high network life time with nearly perfect event report record at 99% without and 96% with clustering. The difference was lost in packet collisions, mostly at the relay nodes, and can be further avoided by aggregating data into bigger packets.

Table 1 Average relative battery levels in each of the circle at the end of network lifetime.

	Central node	First circle (R/2)	Second circle (R)	Thirdcircle (3/2 R)	Fourth circle (2R)
Flood routing	65%	33%	30%	17%	8%
Flood + clustering	78%	32%	35%	15%	8%
Shotgun routing	46%	10%	17%	46%	71%
Shotgun + clustering	53%	14%	35%	57%	76%
SLAM routing	59%	50%	76%	85%	88%
SLAM + clustering	68%	54%	53%	56%	60%
Greedy forwarding	55%	19%	30%	41%	45%
Greedy + clustering	43%	7%	23%	25%	30%
Optimal multi-hop	53%	30%	43%	47%	49%
Optimal + clustering	38%	16%	25%	25%	26%

The third column shows the amount of energy used per successfully re-ported event. Flood routing and shotgun routing – both a very basic routing protocols, performed poorly, requiring an order or even two orders of mag-nitude more power per reported event compared to the more advanced routing schemes. SLAM routing had good energy consumption rate, especially with clustering as it greatly reduced the load off the nodes close to the central base station. However, better results were achieved with greedy forwarding

Table 2 Network parameters at the end of the simulation for each of the routing algorithms.

	Network lifetime	Reported events	Energy used per event	Interference levels
Flood routing	55s	166 (30%)	900 mJ	-68 dBW
Flood + clustering	56 s	28 (5%)	5240 mJ	-68 dBW
Shotgun routing	79 s	341 (43%)	400 mJ	-68 dBW
Shotgun + clustering	78 s	345 (44%)	330 mJ	-68 dBW
SLAM routing	99 s	869 (87%)	53 mJ	-72 dBW
SLAM + clustering	319 s	2385 (74%)	37 mJ	-70 dBW
Greedy forwarding	470 s	4548 (96%)	28 mJ	-68 dBW
Greedy + clustering	553 s	5227 (94%)	29 mJ	-68 dBW
Optimal multi-hop	479 s	4767 (99%)	23 mJ	-75 dBW
Optimal + clustering	645 s	6228 (96%)	24 mJ	-77 dBW

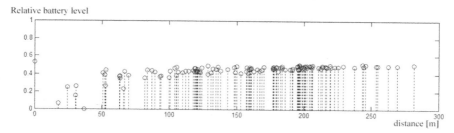

Figure 4 Battery levels at the end of simulation using optimal multi-hop routing without clustering. Nodes with shorter distance from the central node lose battery power faster, resulting in network outage while most of the batteries are still around 50%.

and even more so with optimal multi-hop, requiring only 23 mJ per reported event (1 kb of relayed data).

Finally, the fifth column shows the average interference level through the area covered by network, an important information if the network is to operate below a certain treshold so as not to disrupt existing communication networks within the area, or to stay below detection levels during a conflict. While most routing protocols had similar interference levels, optimal multi-hop achieved a 7–9 dB lower level of interference through the network area, a significant improvement.

Figure 4 shows node relative battery levels for optimal multi-hop routing at the end of network life plotted against the distance from the central base station. Without clustering the nodes closer to the central base station lose energy faster, resulting in much lower battery levels and inevitable drop of nodes from the network. Nodes within the first 50 m from the central base station all had under 30% of battery energy left, while those farther than 50 m (approximately one third of the maximum transmission range) still had on average around 50% battery energy left.

Figure 5 shows the same routing protocol combined with node clustering. A much lower overal battery level can be observed through the network at the end of its lifetime. The node that caused the dropout for this case was actually at a distance of 160 m, practically at the maximum transmission range from the central base station. The few nodes with high battery levels positioned around 100 and 140 m are in fact the relay nodes which start with three times higher battery capacity and levels which they need in order to handle the increased amount of traffic occuring over them. The average node in the network had only around 15–30% energy left in its battery, practically avoiding the energy hole problem.

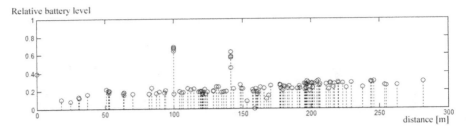

Figure 5 Battery levels at the end of simulation using clustering with optimal multi-hop routing. Remaining battery levels in the network are much lower when compared to the case without clustering, averaging at 15–30%. Several peaks in remaining battery power occuring around 100 and 140 m are from relay nodes which start with three times higher battery capacity.

5 Conclusion

In this paper we addressed the energy hole problem of uneven energy consumption in large scale wireless sensor networks. Several common routing protocols were compared with the advanced smart routing and multi-hop optimization in regards to network lifetime, efficiency of energy use, interference and battery consumption. Impact of clustering on each routing scheme and on reduction of the energy hole problem was considered.

The best results with and without clustering were obtained by optimal multi-hop routing which calculates the best direct path towards the target node and then optimizes it for multi-hop transmission. It achieved a network lifetime of 645 s (20% higher from greedy forwarding) with an average useful data rate of 50 kBps relayed over distance of up to 300 m and only 1 J starting battery power (average energy use of 24 mJ per each 1kb packet). It had an even energy consumption throughout the whole network and practically avoided the energy hole problem.

References

[1] I. F. Akyildiz, W. Su, Y. Sankarasubramaniam, and E. Cyirci. Wireless sensor networks: A survey. Computer Networks, 38(4):393–422, 2002.
[2] D. Estrin, D. Culler, and K. Pister. Connecting the physical world with pervasive networks. IEEE Pervasive Computing, 1(1):59–69, 2002.
[3] L. Srivastava and D. Zrno. AGE@HOME: Radio-enabled environments for independent ageing. Wireless Personal Communications, 51(4), 761–791, 2009.
[4] D. Zrno, D. Šimunić, and R. Prasad. Cooperative indoor radio environment mapping in ad-hoc wireless cognitive networks. Wireless Personal Communications, 2012.

[5] D. M. Blough and P. Santi. Investigating upper bounds on network lifetime extension for cell-based energy conservation techniques in stationary ad hoc networks. In Proc. of IEEE/ACM MobiCom 2002.

[6] E. J. Duarte-Melo and M. Liu. Analysis of energy consumption and lifetime of heterogeneous wireless sensor networks. In Proc. of IEEE GlobeCom, 2002.

[7] M. Bhardwaj and A. P. Chandrakasan. Bounding the lifetime of sensor networks via optimal role assignments. In Proc. of IEEE InfoCom, 2002.

[8] E. J. Duarte-Melo and M. Liu. Data-gathering wireless sensor networks: organization and capacity. Computer Networks (COMNET) Special Issue on Wireless Sensor Networks, 43(4):519–537, 2003.

[9] D. Marco, E. J. Duarte-Melo, M. Liu, and D. L. Neuhoff. On the many-to-one transport capacity of a dense wireless sensor network and the compressibility of its data. In IPSN 2003, LNCS 2634, pp. 1–16, 2003.

[10] J. Li and P. Mohapatra. An analytical model on the energy hole problem in many-to-one sensor networks. In Proc. of IEEE VTC, Fall 2005.

[11] W. R. Heinzelman, A. Chandrakasan, and H. Balakrishnan. Energy efficient communication protocol for wireless micro sensor networks. In Proc. of IEEE Hawaii Int. Conf. on System Sciences, Jan. 2000.

[12] S. Lindsey and C. Raghavendra. PEGASIS: Power-efficient gathering in sensor information systems. In Proc. of IEEE Aerospace Conference, March 2002.

[13] E. M. Royer and C.-K. Toh. A review of current routing protocols for ad-hoc mobile wireless networks. IEEE Personal Communications, April:46–55, 1999.

[14] D. Zrno, D. Šimunić, and R. Prasad. Optimizing cognitive ad-hoc wireless networks for green communications. Journal of Green Engineering, 1(2):209–227, 2011.

[15] D. Zrno. Location awareness in optimization of wireless sensor and cognitive communication networks in indoor environments. Doctoral Thesis, 2010.

[16] G. Martin. An evaluation of ad-hoc routing protocols for wireless sensor networks. Master's Thesis, University of Newcastle upon Tyne, May 2004.

Biographies

Damir Zrno received his dipl.ing. degree in electrical engineering from the University of Zagreb, Croatia in 2002, and his joint Ph.D degree from both University of Zagreb, Croatia and Aalborg University, Denmark in 2010. Since 2002, he has worked at Faculty of Electrical Engineering and Computing in Zagreb, Croatia as a research assistant on several projects in the area of radio communications including network planning, SAR measurements, radio propagation modeling, radio localization, cognitive network simulation and optimization.

Dina Simunic is a full professor at the University of Zagreb, Faculty of Electrical Engineering and Computing in Zagreb, Croatia. She graduated in 1995 from the University of Technology in Graz, Austria. In 1997 she was a visiting professor in Wandel & Goltermann Research Laboratory in Germany, as well as in Motorola Inc., Florida Corporate Electromagnetics Laboratory, USA, where she worked on measurement techniques, later on applied in IEEE Standard. In 2003 she was a collaborator of USA FDA on scientific project of medical interference. Dr. Simunic is a IEEE Senior Member, and acts as a reviewer of *IEEE Transactions on Microwave Theory and Techniques, Biomedical Engineering and Bioelectromagnetics, JOSE*, and as a reviewer of many papers on various scientific conferences (e.g., IEEE on Electromagnetic Compatibility). She was a reviewer of Belgian and Dutch Government scientific projects, of the EU FP programs, as well as of COST-ICT and COST-TDP actions. She is author or co-author of approximately 100 publications in various journals and books, as well as her student text for wireless communications, entitled: *Microwave Communications Basics*. She is co-editor of the book *Towards Green ICT*, published in 2010. She is also editor-in-chief of the *Journal of Green Engineering*. Her research work comprises electromagnetic fields dosimetry, wireless communications theory and its various applications (e.g., in intelligent transport systems, body area networks, crisis management, security, green communications). She serves as Chair of the "Standards in Telecommunications" at Croatian Standardization Institute. She servers as a member of the core group of Erasmus Mundus "Mobility for Life".

Ramjee Prasad (R) is currently the Director of the Center for TeleInfrastruktur (CTIF) at Aalborg University (AAU), Denmark and Professor, Wireless Information Multimedia Communication Chair. He is the Founding Chairman of the Global ICT Standardisation Forum for India (GISFI: www.gisfi.org) established in 2009. GISFI has the purpose of increasing the collaboration between European, Indian, Japanese, North-American, and other worldwide standardization activities in the area of Information and Communication Technology (ICT) and related application areas. He was the Founding Chairman of the HERMES Partnership – a network of leading independent European research centres established in 1997, of which he is now the Honorary Chair.

Ramjee Prasad is the founding editor-in-chief of the Springer *International Journal on Wireless Personal Communications*. He is a member of the editorial board of several other renowned international journals, including those of River Publishers. He is a member of the Steering, Advisory, and Technical Program committees of many renowned annual international conferences, including Wireless Personal Multimedia Communications Symposium (WPMC) and Wireless VITAE. He is a Fellow of the Institute of Electrical and Electronic Engineers (IEEE), USA, the Institution of Electronics and Telecommunications Engineers (IETE), India, the Institution of Engineering and Technology (IET), UK, and a member of the Netherlands Electronics and Radio Society (NERG) and the Danish Engineering Society (IDA). He is also a Knight ("Ridder") of the Order of Dannebrog (2010), a distinguishment awarded by the Queen of Denmark.

Designing Education Process in an Elementary School for Mobile Phone Literacy

Atsushi Ito[1], Yuko Hiramatsu[2], Fumie Shimada[3] and Fumihiro Sato[4]

[1]KDDI Research and Development Laboratories, Japan
[2]Chuo University, Japan
[3]Kamiichibukata Elementary School, Japan
[4]Chuo University, Japan
e-mail: at-itou@kddi.com, susana_y@tamacc.chuo-u.ac.jp,
d353801@city.hachioji.tokyo.jp, fsato@tamacc.chuo-u.ac.jp

Received 22 February 2013; Accepted 27 March 2013

Abstract

In Japan, 53% of children from 10 years old to 17 years old use mobile phones. Filtering technology and systems to protect children are progressing day by day, however, they are still liable to become victims of crime through community sites and 1,239 trouble cases were published in 2010. It is strongly required that children should learn how to treat information on mobile phones. In this paper, we discuss our learning model to learn literacy of mobile phone and the result of our experiment to learn literacy of mobile phone using our model and tool. We call the tool PNS (Pupils Network System) that is designed based on our survey at an elementary school in Tokyo for four years. We confirmed that our methodology is useful to learn literacy of mobile phone.

Keywords: Mobile phone, digital literacy, moral education, e-learning, SNS.

1 Introduction

Mobile communication gives us an access to others whenever wherever we want. We have many benefits from the use of ICT technology. However, from the view point of human developmental psychology children cannot learn this communication of their own accord. A child develops step by step. At first a baby comes to recognize his or her mother and then comes to know its surroundings. After recognizing actual objects, human recognize abstract ones. ICT communication, including mobile communication, doesnt have such steps rising above real places. Operations of mobile phones is so easy that children can use them without lessons even though they cannot recognize what is happened beyond the screen. As a result of such unconscious use cases children become victims of the Internet crimes. The cases of troubles were 1,239 in 2010 in Japan and increased 9.1% compared with the previous year "the Metropolitan Police Department Home Page, February 17, 2011" [1]. The increase of smart phones users in many countries will create the same problems as Japan has. We have studied methods to solve this problem empirically with teachers of elementary schools. In order to teach ICT literacy, we have had trial lessons with 3rd to 6th grade children in Tokyo. Children have opportunity to use PCs in their school. However, they tend to investigate some bookish information or simply learn how to operate a PC. We suggest that it is important for children to learn basic information flow and recognizing human beyond the PC screen, in order to live actively in the ICT society nowadays where abundant information is circulated. In this paper, we discuss our learning model to learn literacy of mobile phone and the result of our experiment to learn literacy of mobile phone using our model and our tool. We call the tool PNS (Pupils Network System) that is designed based on our survey at an elementary school in Tokyo for four years. We confirmed that our methodology is useful to learn literacy of mobile phone. In Section 2, we introduce related works. In Section 3, we discuss our stepwise approach to learn literacy for mobile phone. Then in Section 4, we describe in detail of how we design PNS. Section 5 explains the result of experiment using PNS in 2011. In Section 6, we discuss the result of the experiment. Finally, we present the conclusion of this paper in Section 7.

2 Related Works

The effectiveness of ICT equipment for children was studied practically by using tablet PCs, or PDAs [2] and Interactive white boards [3]. Some teachers created soft ware and shared them. Some schoolbook publishing companies already started ICT contents services for teachers of elementary schools [4]. However, many researches of mobile learning are targeting university children [5, 6]. In 2009, Ministry of Education, Culture, Sports, Science and Technology (MEXT) published a notification to prohibit taking mobile phones to elementary schools [7]. Before the notification, there were several researches on applying mobile phones to education [8, 9].

3 Stepwise Approach of Literacy Education of Mobile Phones

3.1 Environment of Education of Mobile Phones

Since 2001, after e-Japan project of Japanese government, there were many researches to support collaboration learning using mobile phones with built-in camera [10]. After January 2009, MEXT prohibited to bring mobile phones to elementary school because of increasing of crimes, relating to mobile phones, targeting children in elementary school [7]. The notification of MEXT mentioned that cyber crimes are increasing, especially caused by malicious use of mobile phones. Consequently further researches of mobile learning at elementary school cannot be continued. However, there are over 1 million subscribers in Japan. It means 87.8% of Japanese have their own mobile phones according to the Internal Affairs Ministry of research (March 2011) [11]. After that point, researches relating to education using mobile phones were almost dying. However, this situation is not good for children since they have to learn literacy of mobile phone before they use mobile phones actively after they enter junior high school. However, in the guideline about the information of the education in Japan [12], MEXT wrote that

> When teaching subjects etc., each school should improve learning activities so that pupils become familiar with information devices, such as computers and information and communications networks, acquire basic operation skills, such as typing letters on a computer keyboard, and information ethics and are able to use information devices appropriately. In addition to these information devices, each school should also use other teaching materials and aids properly, such as audiovisual materials and teaching and learning devices.

The prohibition to use mobile phones in a elementary school and the guide line are contradictory. So that, we have to teach children how to use mobile phones and mobile information. We started study of literacy learning of mobile phones since 2009 [13–16]. Our target is how to make children in elementary school understand "invisible" flow of information in the cyber world. A mobile phone is a kind of gateway to access the Internet. We set three directions to teach mobile phone literacy.

- Mobile phone is a personal device, so that it is not easy that their parents observe every behavior of children. There is a possibility for a child to be a sacrifice of crime on mobile phones. It is important for them to learn literacy of mobile phone.
- Mobile phone is a small and light, so that is way for children to use. Also, a mobile phone equips a lot of useful devices and functions such as camera, recording sound and input text.
- It is easy to send data, such as photo or mail, from a mobile phone to other phones by using IR (Infra Red) and Bluetooth by using mobile phones. They can realize and understand how the information flows in the network.

3.2 Our Learning Model for Literacy of Mobile Phones

In Figure 1, our model for education of literacy of mobile phone is described. There are two main parts, one is developing skills and another is moral. In the skill part, there are six boxes. Each box means each grade of an elementary school and displays the step-by-step learning of functions of mobile phones and lessons in a class such as gathering materials when they visit out side of the school. We have not yet performed mobile phone literacy (skill up) class for 1st, 2nd and 6th grade, so that the box is described by dotted line. Details of classes for 3rd, 4th and 5th grade will be explained later. On the other hand, right part is moral education such as not to expose face clearly, not to write abuse, not to expose personal information in profile site. Moral education is important to prevent risk in the Internet and manner. Moral education is simple, but it is important to learn it again and again until the knowledge is a part of a brain. Firstly, we define the following flow to learn moral of mobile phone as described in Figure 2. As an introduction, children attend a lecture of general information of risks of mobile phones, then, they learn about what is personal information and importance of personal information. Next, learn about what is communication and at last, they learn about the crimes especially contact from unknown person. As skills of using mobile phones, we

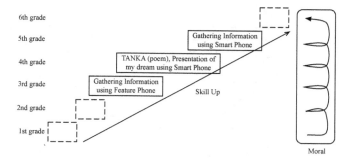

Figure 1 Learning model for literacy of mobile phones.

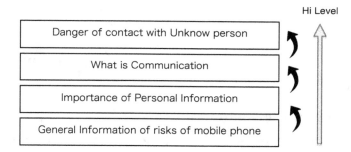

Figure 2 Flow of learning moral of mobile phones.

define the following flow described in Figure 3. This flow is based on the main functions of mobile phones. A mobile phone has five major functions as follows.

1. Communication (voice and data)
2. Application
3. Camera
4. Address book
5. SNS

A flow of learning skill of mobile phones is based on these core functions. Firstly, taking picture and input text. Then write articles using picture and text. Next, they study what is communication by sending picture to others. Then provide information to others and how to respond to others. We started literacy education of mobile phones based on this model and continuing from 2009. On 2010, we apply this model to fifth grade of the elementary school

Skill Level (Hight)

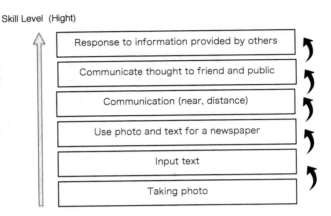

Figure 3 Flow of learning skill of using mobile phones.

and 2011, we continued the education to fourth grade children. We would like to explain details of our trial.

3.3 Learning Trial in Third Grade

As a lesson of moral, we learned general information of risks of using mobile phones. We used one unit (45 minutes) for this purpose. Then we learned the risks again by using mobile phones. The purpose was to learn information sharing and how to manage risks relating to sending information. For that purpose, we used mobile phones without SIM (Subscriber Identity Module) cards to prevent unexpected communication to outside a classroom. Using this mobile phone, they learned how to send and receive photos using IR. Figure 4 displays the three steps of learning communication using mobile phones. At first, bottom of the figure, they learned face-to-face communication since IR communication is limited within 30 cm. Then they created their own newspaper based on the research at wholesale market and the observation of trees in around the school. They are satisfied with the result and felt as if they developed a newspaper as real one using photos and text gathered by mobile phones. The flexibility and portability of mobile phones were useful and effective. Also, children who are not aggressive in a class could also join the development process easily.

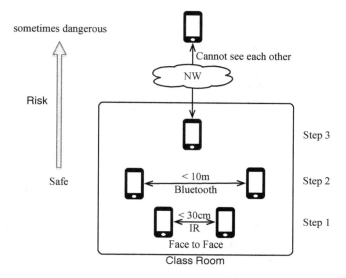

Figure 4 Three steps to learn communication.

3.4 Learning Trial in Fourth Grade

3.4.1 Learn about Communication

In 2011, we firstly performed moral class for understanding how fast information is superadded in the Internet and how dangerous it is. We also used mobile phones without SIM card and send a photo each other by using IR. We measured how fast the photo is shared in the class. It was only three minutes. Then we used Bluetooth to send photo each other as described in the middle of Figure 4. This is not face-to-face communication, so that this is a preparation of real Internet communication. After these trials, we learned the crime case of young schoolgirl caused by mobile phones and understood how to protect themselves from malicious people.

3.4.2 Learn about Message Exchange

In July 2011, we performed three classes, total 90 children, to learn Japanese short poem, TANKA [17], by using real Internet. TANKA has traditional message exchange format in Japan and it started about 1,000 years ago. TANKA is similar to time line, however, it is very polite communication tool for ancient Japanese people. One student creates short sentence (17 characters), then one of classmate adds short comment as displayed in Figure 5. We

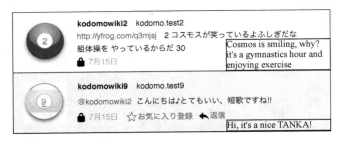

Figure 5 TANKA (top) and Reply (bottom).

developed a closed group in Twitter. Details were described in our previous paper [18]. Children heard about risks of the Internet, however, they could not understand the risks clearly. This is the reason that we prepared closed group in Twitter. In this environment, only user ID and articles can be accessible. They learned how to describe articles on the Internet and comment on it. They knew about the Internet and risks on the Internet, however, most of them did not have experiences to use it. Their knowledge came from their parents, TV and books. They were very excited to use mobile phones and the Internet. When teachers asked them to take pictures, 28 of 30 children in a class took face of their friends. They almost forgot what they learned last year at the moral of using mobile phones. The teacher asked them to remember what they had learned. Then they remembered that it was not allowed to take photo of face of their friends and be careful to submit articles on the Internet. After preparation described above, they took photos and tried to create TANKA relating to the photo. They worked aggressively. Some children created several TANKA. Of course, they made a lot of mistakes and uploaded TANKA before completion. After each student uploaded TANKA, other children commented on them. They worked very well. They could add gentle and thoughtful reply to TANKA. Figure 5 describes this process. First, #2 sent TANKA, and then #9 commented on this TANKA. In the classroom, they shared timeline by using projector and presented their TANKA and printed it. Figure 6 shows the result of a questionnaire to children who attended this trial. The questionnaire asked whether they would like to use the smart phones in the class again or not. At least 70% of children would like to use it again. However, 300% of children answered negative since text input on mobile phones was not easy for them.

We think that the difficulty of inputting text on mobile phones did not neglect their motivation. Also, projection on screen to share the contents and

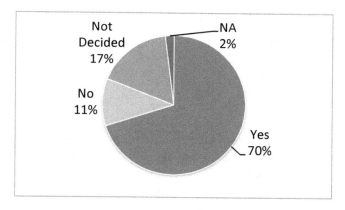

Figure 6 Children's preference to use mobile phone again in a class.

printing to personalize their TANKA work increased their motivation very much. On the other hand, some children uploaded 20 photos. We thought that the environment to study skill and moral of mobile phones are required to increase their literacy. We think that this kind of environment to allow children to represent their feeling directly is very valuable for them. So that we believe that the idea of PNS should work well. In a class, the teacher checked children's work very carefully. About half of the children in the class answered the questionnaire that the operation of smart phone was difficult, however, they would like to use the smart phones in the class again as the same percentage as in other classes. We can summarize our survey that young children in an elementary school cannot imagine the worst case and accept the fact directly. So that, they are sometimes caught by malicious trap. In the next section, we define the requirements for a system to support ethic education for the mobile phone literacy.

4 Designing PNS: A Tool for Ethic Education of How to Use Mobile Phones

Now we would like to start to design PNS (Pupils Network System) for ethic education of mobile phone literacy. We intended to use PNS from third grade in Figure 1. With this system, children upload information and communicate with others. The most remarkable point is that this site will give a place for children to evaluate the subject with autonomy and check each other. When a student sends some information, the other children in his group check the

Table 1 Functions of each player.

Target Recommendation	Suggestion and Write articles	Check details	Approval	Write Comments	Perusal
Student (contributor)		○		○	○
Student (advisers)	○	○	○		○
Teacher	○	○	○	○	○
Parents					○

posted articles and send back an evaluation report. If two or more persons agree, the article will be uploaded on the site. Children in 3rd to 6th grades can vote for good articles (refer to Table 1). The student who gets good feedback can be the new leader of a group. Teachers and parents can peruse articles. As a result we wish to work on the following points:

- From the results of the moral class – Feeling of reality, taking responsibility for what the student writes.
- From the view of internet literacy education – Safety, including communication with distant places.

4.1 Behavior of PNS

First of all, we would like to explain the behavior of PNS. Figure 7 displays the flow of the submission to PSN and publishing articles to parents.

Step 1: A student inputs an article (maybe with photo).

Step 2: Firstly, the article should be checked by three children. They are classmates and/or children in senior classes. They checked that the article is appropriate to submitted to PNS, such as the theme is relating to subjects and there are no grammatical mistakes. Also, they check there is no prohibited word in the articles. At this step, children learn ethics of using mobile phones and the Internet services.

Step 3: Then the article is opened to other children. They can add comments on that article.

Step 4: A teacher checked the article and if it is good to open to parents, he/she uploads the article to the website of the school.

Step 5: Parents can see the article, such as reports of outdoor activity or poems.

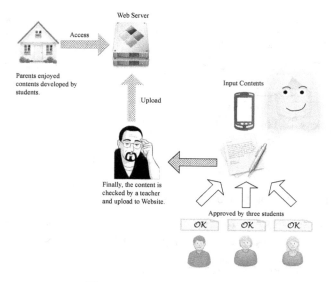

Figure 7 Outline of dataflow of PNS.

4.2 Requirements for PNS

Then, we set up the following requirements for PNS as a tool for ethic education.

(R1) Write articles and add comments on that article easily.

(R2) Approval process should be prepared. Children sometimes write inappropriate article to be published.

(R3) The article should be described in not too long but not too short. Teachers can decide the length.

(R4) children sometimes use bad words, so the system should automatically check such words and displays alert.

(R5) PNS should be closed in a school. Going outside and accessing from outside should be prohibited to protect personal information of children to prevent criminal.

(R6) Every grade, classmates are changed, so to reduce operation by teacher, user database of PNS should be automatically updated by reading a name list of a class.

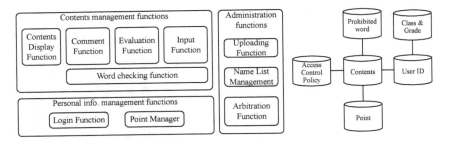

Figure 8 Functions and its data model of PNS.

(R7) PNS should be controlled by teachers in detail.

(R8) The UI of PNS should be beautiful and considered to meet the recognition process of young children under 10 years old.

(R9) Encourage children to access the system.

(R10) Easy operation. Teachers are busy, so maintenance free is ideal for them.

(R11) The operation cost should be minimized. Free to use is ideal.

Twitter has good appearance, free to use and is stable. But there is no administration and keeping safety function in Twitter, so we thought that we have to develop PNS by ourselves.

4.3 System Architecture of PNS

Based on the above discussions, we designed the PNS. By introducing simple and easy UI of Twitter, we would like to design several functions to protect children. Figure 8 displays functions of PNS and data types that are used in PNS. There are three components in the system, Contents management functions, Administration functions and Personal information management functions. Each function should satisfy requirements R1-9. R10 and R11 are general requirements, so that they are not directly implemented. The outline of each component is as follows:

Contents management functions:

- Input Function: Each student can input articles to PNS through mobile phones. (R1) (R3)

- Evaluation Function: Who assigned as an evaluator for a student can review and evaluate that article. If three reviewers agree, that article can be published to all children. If the article is a question, they give an answer on that article. (R2)
- Comment Function: Any child can comment on an article. (R1)
- Word checking function: When inputting article, the system checks the words in that article include prohibited words, such as kill, foolish, or not. If there is a prohibited word, the system displays alert. (R4)
- Contents Display Function: Each article is displayed in a readable form, color and font size. (R8)

Administration functions:

- Uploading Function: A teacher can upload selected article to the website of a school. (R7)
- Name List Management: A teacher can manage ID of student simply relating to name list of a class. (R6)
- Arbitration Function: If some troubles happen on PNS among children, a teacher can interrupt the thread. (R7)

Personal information management functions:

- Login Function: This function allows each student to enter the system and only the student is allowed to enter. (R5)
- Point Manager: The number of comments is reflected as points of an author of an article. (R9)

Figure 9 displays the UI of PNS on Android.

5 Experiment of PNS in 2011

We used PNS at February 2012 in fourth grade classes. They input their photo when they are infants and describe their dream, such as to be a baseball player, to be a doctor. At last, all the articles are projected in the gymnasium and shared by children and parents. It was easy for children to use, they soon understand how to input articles and submit articles. They were satisfied to complete articles by themselves. We asked them what is the difference between Twitter and PNS to the third class of the fourth grade (30 children each). They answered on the questioner sheets. When we asked them about common point between Twitter and PNS, most popular answer was Upload file and Input Text. Only 6 of 30 children answered that both systems use Internet. Next, we asked them about difference between them. The most pop-

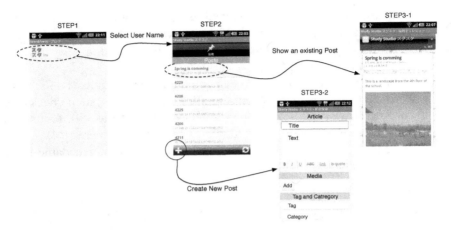

Figure 9 User interface of PNS.

Table 2 Important points when we use the Internet from a mobile phone (including multiple answers).

Answer	Number of Ans.
Consider to handle personal information	14
Do not write real name (including friends)	14
Do not upload photo of persons	10
Do not access unknown sites	9
When using Twitter, it is required to ask parents	2

ular answer was that Twitter was used for TANKA study and PNS was used for describing their dreams. In their mind, only the difference of contents was remaining. We encouraged them to remember the moral class last year. They found that they violated rules to prevent risks to use the Internet such as uploaded photo with face. Also, no one could not aware that difference of management mechanisms of both systems and of users. Even they learned about moral last year, they forgot details of the lesson. In Table 2, the important points when we use the Internet form a mobile phone (including multiple answer) are listed up. Also, impression of the Internet literacy class was described in Table 3. Their answers were interesting. They answered both under positive phycology and passive points.

Table 3 Impression of Internet literacy class.

Answer	Number of Answers
Do not forget risks of the Internet	7
Be careful when using the Internet	5
Fun and attractive	2
Understand what is prohibited and allowed	2

6 Discussion

One of the major SNS in Japan is Mixi [19]. A report said that about 90% of teenager users of Mixi uses that service from mobile phones. It is clear that children of elementary school may access services on mobile phones soon after entering junior high school. It is important that they should learn and do practice how to use mobile phones safely through actual use of mobile phones under the controlled situation such as in a class room in the elementary school. As the result of experiments, children in an elementary school have knowledge of the Internet, however, their reaction to the problems is not always correct. Also their knowledge is not well organized, so if they do carefully upload their contents using mobile phones, they sometimes make mistakes. It is true that learning becomes real knowledge through mistakes. We confirmed that PNS is useful for them to learn literacy of mobile phones by making a lot of mistakes. We would like to give them skills to see facts using critical power. They can learn the critical power to the information on the Internet through learning and mistakes. It is also important to compare facts to cultivate criticism for the information on the Internet. We consider that PNS is useful to learn literacy of mobile phone as an SNS in a school. We would like to add moral lessons on using ICT equipment in PNS to cultivate literacy of using mobile phone and the Internet. One student who attended these classes did not understand the risk of accessing Internet. It is safe to use the Internet in an elementary school, however, there is a possibility to access the Internet from his home and becomes a sacrifice of a crime. We can teach him the risk of the Internet access through the individual guidance when teachers fond the problem of understanding level of risks of the Internet in a classes. Also, there is a possibility to apply education using mobile phones to change or modify the structure of their cognitive functioning to adapt to the changing demands in a class as described in [20]. Learning using a mobile phone is good for personalized learning. In usual class, some children understand well and others are not. However, learning using a mobile phone allows a student to learn things step by step by confirming steps.

6.1 Further Study

To learn functions of mobile phones is easily to meet the step-by-step method as mathematics, etc., since children are familiar with accessing ICT equipment as a game machine or digital camera. It is natural for them to use mobile phones as Digital Native. However, the lesson of moral of using ICT equipment like mobile phones is a little bit different. Of course, it is not difficult as mathematics or foreign languages, but it is easy to forget important issues relating to moral on the Internet, such as not to upload photos that include faces of their friends, not to disclose their email addresses and phone numbers, etc. It is important to learn again and again especially on the moral of using mobile phones. We think that it is important to take learning history of moral education including attitude and response in the class and to share that history from the first grade to sixth grade. Also it is important to construct a system to teach moral based on the history of the children. We think that PNS may have possibility to realize that function. PNS is mainly covers left part (skill education) in our education model described in Figure 1, so that we would like to add function on PNS that realizes right part (moral education) of Figure 1.

7 Conclusion

Taking conclusion of research, we made educational SNS application for children to learn about communication literacy of the Internet by considering security issue based on our learning model. Our PNS had has some steps and it took time to circulate articles. However, convenience was not our purpose. We made the children to consider about the subjects. By using of PNS, children could consider and learn both good points and the risks of Internet communication step by step. In addition we would like to use this application not only for mastering operation of mobile phones, but also to connect with others in the society they belong to. There may be many other fields to help children connect to others, using PNS. We continue to do research in elementary schools.

Acknowledgments

We would like to thank the teachers, Mr. Sato (the principal), Mr. Nishioka (the ex-principal), Ms. Tanaka, Ms. Konda, Mr. Oba, Ms. Tokumasu, Mr. Yamada, Mr. Nakamura, and the children in Kamiichibukata Elementary

School for their performance in this research. Furthermore, we thank the volunteers from Professor Sato's Laboratory of Chuo University for their support of this research.

References

[1] HP of Tokyo Metropolitan Police Department, http://is702.jp/special/332/#netiquette-t1.

[2] T. Ishizuka et al. Practical study on elementary school class with PDAs. In Studies in Informatics, pp. 5–14. Shizuoka University, 2003.

[3] Y. Nakahashi. Instructional strategies for promoting joint thinking and dialogue among learners by means of interactive whiteboards. Japan Journal of Educational Technology, 33:373–382, 2010.

[4] Tosyo e-net, http://ten.tokyo-shoseki.co.jp/.

[5] M. Kuroki et al. An educational practice of the active learning in the university class. *Japan Society of Educational Information*, pp. 190–193, 2010.

[6] C-C. Tsai et al. (Eds.). Special reports on the future education opened up by mobile learning. *Computer & Education*, 28, 2010.

[7] Notification to prohibit taking mobile phones to elementary schools, http://www.mext.go.jp/b_menu/hakusho/nc/1234695.htm.

[8] M. Ohkubo et al. Development and evaluation of a system supporting collaborative learning using camera-equipped mobile phones., Japan Society for Educational Technology, 28:189–192, 2004.

[9] K. Ito et al. Research on the function of mobile communication system to support outdoor study. Japan Association for Communication, Information and Society, 1(2):12–15, 2005.

[10] The Cabinet Office. The interim report about the environmental maintenance that the young people use internet safety in Japan. Planning Interactive Explanations, 2011.

[11] The Internal Affairs Ministry, White Paper, Information and Communications in Japan, p. 88, March 2010.

[12] Ministry of Education, Culture, Sports, Science and Technology. The Guideline about the Information of the Education in Japan, 2010.

[13] Y. Hiramatsu et al. Information Literacy in out-of-class activities. Council for Improvement of Education through Computers, 2009.

[14] Y. Hiramatsu et al. Effects of using mobile phones in primary education. Japan Association for Communication, Information and Society, pp. 2–3, 2010.

[15] Y. Hiramatsu et al. A report on teaching practical usage of mobile phone in elementary school. Computer & Education, 29:76–79, 2010.

[16] Y. Hiramatsu et al. A study of teaching digital literacy for children: Moral education to use the internet on mobile phones. In Proceedings of the IADIS International Conference e-Learning, pages 337–340, 2011.

[17] http://en.wikipedia.org/wiki/Waka_(poetry)#Tanka.

[18] A. Ito et al. Designing PNS (Pupils Network System) for ethic education of mobile phone literacy. In Proceedings of the IB2COM, 2011

[19] http://mixi.jp.

[20] R. Feuerstein, S. Feuerstein, L. Falik, and Y. Rand. Dynamic Assessments of Cognitive Modifiability. ICELP Press, Jerusalem, Israel, 1979/2002.

Biographies

Atsushi Ito received B.S. and M.S. degrees from Nagoya University in 1981 and 1983, respectively. He also received a Ph.D. from Hiroshima City University in 2007. He is currently with KDDI R&D Laboratories. During 1991–1992, he was a visiting scholar at the Center for the Study of Language and Information (CSLI) of Stanford University. His current research interests include open platform for mobile communications and applications of ICT for education and healthcare.

Yuko Hiramatsu received B.A. and M.A. from Sophia University in 1979 and 1984 respectively. After R&D Initiatives duty, she is now a visiting lecturer of Faculty of Economics at Chuo University. She received the best paper award from CEIC (Community for Innovation of Education and learning through computers and communication networks) in 2010. Her research interest includes education method of information literacy.

Fumie Shimada received B.Ed. degree from Niigata University in 1985 and immediately became a teacher of elementary school. She became a vice principal at Kamiichibukata elementary school in Tokyo in 2009. She received the best paper award from CEIC (Community for Innovation of Education and learning through Computers and communication networks) in 2010. Her research area includes applying ICT for primary education.

Fumihiro Sato graduated from Waseda University in 1974 (B.Ed). He is a professor at Chuo University's Faculty of Economics specializing in ICT. Prior to this, he worked with the Japan Information Processing Development Corporation (JIPDEC) for 20 years. During 2002–2004, he was a visiting scholar at the Center for Design Research (CDR) of Stanford University. His current research interests focuses on the tasks of ICT for education in the South East Region.

Adaptive Maintenance Optimization Using Initial Reliability Estimates

Khalid Aboura[1] and Johnson I. Agbinya[2]

[1]College of Business Administration, University of Dammam, Saudi Arabia;
e-mail: kaboura@ud.edu.sa
[2]Department of Electronic Engineering, La Trobe University, Kingsbury Drive,
Victoria 3083, Australia; e-mail: j.agbinya@latrobe.edu.au

Received 27 March 2013; Accepted 1 April 2013

Abstract

This paper presents a procedure for determining optimal times for the replacement of a large number of identical items operating under similar conditions. A collective maintenance policy is considered due to the prohibitive cost of individual replacement upon failure. Replacements of all items occur at prescribed points in time, assuming that a reasonable proportion of failed items are tolerated. The first replacement time is chosen using initial reliability estimates. Successive replacement times are determined as failure and survival information is gained. The procedure is developed for the data collection scenarios of complete and interval censored data.

Keywords: Maintenance optimization, block replacement, statistical inference.

1 Introduction

We present an adaptive maintenance optimization solution for the replacement of a large number of identical items operating independently under similar conditions. An example would be the replacement of light bulbs in city streets and large department stores. In a typical situation, individual replacement upon failure is expensive and planned replacement of all items is preferred. A minimization of maintenance costs is essential given the sizes of the structures. Upon the introduction of new equipment or at the start of a study, reliability estimates are often available only in the form of vendor information or informed judgment from maintenance operators. As failure and survival data are collected, a better assessment of the life length characteristics of the items becomes possible, allowing a more effective replacement procedure. Adaptive maintenance strategies apply well in this situation. Mazzuchi and Soyer [13] use an updating scheme in a decision theoretic set up to allow the maintenance policy to adapt to failure information. Earlier maintenance optimization algorithms were developed based on the assumption that the lifetime characteristics of the items are specified at the start of the operations. Zio and Compare [21] give a good perspective on the three main aspects of corrective, preventive and dynamic maintenance. For a survey of maintenance models when the parameters of the failure distribution are unknown but constant see Valdez-Florez and Feldman [18] and Sarkar et al. [14]. When the parameters of the failure distribution are considered random variables, a Bayesian parametric analysis often ensues [20]. In such case, a growing number of models present exceptions to the static nature of the early policies. Some of these work include Fox [9], Bassin [4], Bather [5], Frees and Ruppert [10], Aras et al. [3], Benmerzouga [6], Mazzuchi and Soyer [13] and subsequently Sheu et al. [15, 16], Dayanik and Gurler [7], Juang and Anderson [11], and Flage et al. [8]. The maintenance scenario considered in this paper departs from the traditional block and age replacement scenario. Here, an item under consideration is assumed to remain failed upon failure until the next planned replacement time, when all items are replaced. The manual cost of inspection and replacement of items prohibits an individual maintenance. In addition, we assume that the failure of a reasonable proportion of the items does not bring to stop the whole structure and therefore allows a planned strategy for the simultaneous replacement of all items. In the following, the Weibull distribution is assumed to be the lifetime model. A maintenance optimization solution is outlined and simulated examples are used to illustrate the methodology.

2 The Maintenance Scenario

We consider a structure of M identical items operating independently of each other under similar conditions. At prescribed points in time $T_1, T_2, \ldots$, etc., all items are replaced by new ones. An item that fails before the next replacement time remains failed. We let T_0 be time 0. As failures accumulate between the replacement times, two types of data collection are possible: (1) the exact failure times are recorded (complete data) and (2) the number of failures per time interval is recorded (interval censored data). We treat both cases and consider only the case of the numbers of failures between replacement times in case (2). That is the interval censored data consist of the number of failures in $[T_{i-1}, T_i)$, $i = 1, 2, \ldots$. The more general case of interval censored data involves inspection points in $[T_{i-1}, T_i)$ where the numbers of failures are recorded between the inspection points. The extension to the inspection case is straightforward given that the inspection times are fixed. A further extension would be to consider the inspection times as decision variables in the setting of an optimal maintenance strategy.

3 The Lifetime Distribution

The lifetime of an item, say T, is assumed to have a Weibull probability distribution with reliability function $R(t/\lambda, \beta) = e^{-\lambda t^\beta}$. At time T_0, the prior distribution of the parameters (λ, β) is assessed using available information. Aboura [1] introduces an approach for the construction of a prior distribution for (λ, β) when initial reliability estimates are available. We use the methodology of Aboura [1] in constructing a prior distribution.

3.1 The Prior Joint Distribution

Let T be the lifetime of the item under consideration. In practice, informed judgement about T is often available in the form of vendor information, engineering knowledge or experience in the field. We assume that reliability estimates $r^{(n)} = (r_1, r_2, \ldots, r_n)$ are provided for different mission times. r_i, $i = 1, 2, \ldots, n$, is considered an expert's estimate of the reliability for mission time t_i. Let r_{n+1} be a lower bound so that

$$1 \equiv r_0 \geq r_1 \geq r_2 \geq \ldots \geq r_n \geq r_{n+1} \geq 0, \quad 0 \leq t_1 \leq t_2 \leq \ldots \leq t_n < \infty \tag{1}$$

Assuming a Weibull model for the lifetime T with reliability function $R(t/\lambda, \beta) = e^{-\lambda t^\beta}$, the likelihood model adopted for the expert's data is the Dirichlet distribution [19], where

$$p(r_1, r_2, \ldots, r_n/\lambda, \beta) = \frac{\Gamma(b)}{\prod_{i=1}^{n+1} \Gamma(ba_i)} \frac{\prod_{i=1}^{n+1} (r_{i-1} - r_i)^{ba_i - 1}}{(1 - r_{n+1})^{b-1}} \tag{2}$$

b and $\{a_i\}_{i=1}^{n+1}$ are functions of λ and β, and $\sum_{i=1}^{n+1} a_i = 1$, $b > 0$, $a_i > 0$, $i = 1, \ldots, n + 1$. The marginal distributions are

$$p(r_i/\lambda, \beta) = \frac{\Gamma(b)}{\Gamma(ba_{i*})\Gamma(b(1 - a_{i*}))} \frac{(r_i - r_{n+1})^{b(1-a_{i*})-1}(1 - r_i)^{ba_{i*}-1}}{(1 - r_{n+1})^{b-1}} \tag{3}$$

with $a_{i*} = \sum_{j=1}^{i} a_j$, means $E(r_i/\lambda, \beta) = r_{n+1} + (1 - r_{n+1})(1 - a_{i*})$ and variances $V(r_i/\lambda, \beta) = (1 - r_{n+1})^2 a_{i*}(1 - a_{i*})/(1 + b)$.

Given λ and β, a perfect expert input would be $e^{-\lambda t_i^{\beta}}$ for the reliability at time t_i. Following Lindley [12] and introducing modulation parameters $\mu^{(n)} = (\mu_1, \mu_2, \ldots, \mu_n)$ and $\sigma^{(n)} = (\sigma_1, \sigma_2, \ldots, \sigma_n)$, we assume that $E(r_i/\lambda, \beta) = \mu_i + \sigma_i e^{-\lambda t_i^{\beta}} = e_i$ to obtain $a_{i*} = (1 - e_i)/(1 - r_{n+1})$, $a_i = (e_{i-1} - e_i)/(1 - r_{n+1})$, $i = 1, 2, \ldots, n$, $e_0 = 1$.

To select b, a variety of approaches can be used. The covariance structure of $r^{(n)}$ is imposed by the choice of the Dirichlet model, and for large (small) values of b, small (large) variances result. Aboura [1] determines $b(\lambda, \beta)$ for the two cases where upper and lower bound functions of (λ, β) are provided for the expert's estimates. In practice, it is often hard to obtain information about the spreads of the expert estimates. In that case the variances in the expert likelihood models are set to reasonable values and varied as part of the analysis. In the following, b is set to a constant.

The prior joint density distribution of λ and β, given the expert's informed judgment $r^{(n)}$, $p(\lambda, \beta/r^{(n)})$, is obtained through the application of Bayes theorem:

$$p(\lambda, \beta/r^{(n)}) \propto p(r^{(n)}/\lambda, \beta)p(\lambda, \beta) \tag{4}$$

In practice it is often the case that no prior knowledge exists about λ and β. In the following, we will assume a flat prior for λ and β.

The prior density distribution $p(\lambda, \beta/r^{(n)})$ is obtained numerically. Deriving the posterior distributions for inference requires numerical integration. Soland [17] introduced a mixed prior distribution for the parameters of the Weibull distribution that allows for closed form posterior distributions for failure and right-censored observations. β has a discrete distribution and λ a natural conjugate Gamma distribution. λ and β are assumed independent. Mazzuchi and Soyer [13] used Soland's distribution for the parameters of the Weibull lifetime distribution. A discretized Beta distribution is used for the parameter β. Although such a use of Soland's distribution does provide a starting prior joint density, one could dispute the feasibility of collecting any direct information from an expert about β, the abstract model parameter β not having any physical meaning. One can also argue about the arbitrariness used by Mazzuchi and Soyer [13] to select the range of the discretized Beta distribution for β, unless this range is made to cover most of the likely values of β. Here we remedy these shortfalls by constructing a prior density for (λ, β) using estimates of observables. The range of β and the dependence structure of (λ, β) result naturally from the expert opinion elicitation procedure.

3.2 A Mathematically Tractable Prior

The distribution of Soland [17] is extended to include dependence and fitted through moments to the prior distribution $p(\lambda, \beta/r^{(n)})$. The procedure to construct $g(\lambda, \beta/r^{(n)})$, the fitted distribution, consists of the following four steps:

Step 1. Compute the prior distribution $p(\lambda, \beta/r^{(n)})$.

Step 2. Determine the range of β. Let $[\beta_l, \beta_u]$ be the interval where most probability for β falls, i.e.

$$\text{Prob}(\beta \in [\beta_l, \beta_u]/r^{(n)}) = \int_{\beta_l}^{\beta_u} \int_0^\infty p(\lambda, \beta/r^{(n)}) d\lambda d\beta \cong 1.$$

Step 3. Discretize β. Let k be a chosen number so that $\beta \in \{\beta_1, \beta_2, \ldots, \beta_k\}$, where $\beta_j = \beta_l + (j-1)(\beta_u - \beta_l)/(k-1)$, $j = 1, 2, \ldots, k$.

Step 4. Obtain $g(\lambda, \beta/r^{(n)})$. For $j = 1, 2, \ldots, k$, compute q_j the normalized value for $\text{Prob}(\beta = \beta_j/r^{(n)})$ after discretization

$$q_j = \frac{\int_0^\infty p(\lambda, \beta_j/r^{(n)}) d\lambda}{\sum_{j=1}^k \int_0^\infty p(\lambda, \beta_j/r^{(n)}) d\lambda}) \tag{5}$$

Compute the conditional mean

$$E_j = E(\lambda/\beta_j, r^{(n)}) = \int_0^\infty \lambda p(\lambda/\beta_j, r^{(n)}) d\lambda$$

and the conditional variance of λ, $V_j = V(\lambda/\beta_j, r^{(n)})$. Let c_j and d_j be such that $c_j = E_j^2/V_j$ and $d_j = E_j/V_j$. The fitted conditional prior distribution for λ given β_j and $r^{(n)}$ is Gamma(c_j, d_j), i.e. $g(\lambda/\beta_j, r^{(n)}) = d_j^{c_j} \lambda^{c_j-1} e^{-d_j \lambda}/\Gamma(c_j)$ where $g(\lambda, \beta_j/r^{(n)}) = g(\lambda/\beta_j, r^{(n)}) q_j$, for $j = 1, 2, \ldots, k$. The fitted prior distribution preserves the dependence structure of λ and β contained in the original prior density. By construction it also preserves the conditional first two moments of λ, while doing the same for β if k is large enough.

Using the prior distribution $g(\lambda, \beta/r^{(n)})$, the first replacement time T_1 is determined based on costs and the expected number of failures in $[T_0, T_1)$. Subsequently, at each replacement time T_{i-1}, $i = 2, 3, \ldots$, the next replacement time T_i is determined using costs and the expected number of failures in $[T_{i-1}, T_i)$, posterior to the observed failure and survival data in $[T_0, T_{i-1})$. We begin by deriving the posterior distribution of (λ, β) for the two data collection cases. We then derive the optimal replacement times.

3.3 The Posterior Distribution for Complete Data

Given the expert's input $r^{(n)} = (r_1, r_2, \ldots, r_n)$ for $t^{(n)} = (t_1, t_2, \ldots, t_n)$ and an appropriate choice of b, the posterior distribution of (λ, β) for observed failure times $f^{(m)} = (f_1, f_2, \ldots, f_m)$ and survival times (right censored observations) $s^{(l)} = (s_1, s_2, \ldots, s_l)$ is

$$p(\lambda, \beta/f^{(m)}, s^{(l)}, r^{(n)}) \propto \left\{ \prod_{i=1}^{m} \lambda \beta f_i^{\beta-1} e^{-\lambda f_i^\beta} \right\} \left\{ \prod_{u=1}^{l} e^{-\lambda s_u^\beta} \right\} p(\lambda, \beta/r^{(n)})$$

(6)

$p(\lambda, \beta/r^{(n)})$ is replaced by $g(\lambda, \beta/r^{(n)})$ (with parameters c_j, d_j and q_j) to allow for a closed form of the posterior distribution. For $\beta_j = \beta_l + (j - 1)(\beta_u - \beta_l)/(k - 1)$, $j = 1, 2, \ldots, k$ and $0 \le \lambda < \infty$, the joint posterior distribution follows from (6) with normalizing constant $\sum_{j=1}^{k} \int_0^\infty p(\lambda, \beta_j/f^{(m)}, s^{(l)}, r^{(n)}) d\lambda$ to produce

$$p(\lambda, \beta_j/f^{(m)}, s^{(l)}, r^{(n)}) =$$

$$\frac{\beta_j^m \left\{ \prod_{i=1}^{m} f_i^{\beta_j-1} \right\} d_j^{c_j} q_j \lambda^{m+c_j-1} \exp\left[-\lambda(d_j + \sum_{i=1}^{m} f_i^{\beta_j} + \sum_{u=1}^{l} s_u^{\beta_j}) \right] / \Gamma(c_j)}{\sum_{j=1}^{k} \beta_j^m \left\{ \prod_{i=1}^{m} f_i^{\beta_j-1} \right\} q_j \frac{\Gamma(m+c_j)}{\Gamma(c_j)} \frac{d_j^{c_j}}{\left(d_j + \sum_{i=1}^{m} f_i^{\beta_j} + \sum_{u=1}^{l} s_u^{\beta_j} \right)^{m+c_j}}}$$

(7)

The marginal posterior distribution of β, $\int_0^\infty p(\lambda, \beta_j/f^{(m)}, s^{(l)}, r^{(n)}) d\lambda$, is

$$p(\beta_j/f^{(m)}, s^{(l)}, r^{(n)}) =$$

$$\frac{\beta_j^m \left\{ \prod_{i=1}^{m} f_i^{\beta_j - 1} \right\} q_j \frac{\Gamma(m+c_j)}{\Gamma(c_j)} \frac{d_j^{c_j}}{\left(d_j + \sum_{i=1}^{m} f_i^{\beta_j} + \sum_{u=1}^{l} s_u^{\beta_j} \right)^{m+c_j}}}{\sum_{j=1}^{k} \beta_j^m \left\{ \prod_{i=1}^{m} f_i^{\beta_j - 1} \right\} q_j \frac{\Gamma(m+c_j)}{\Gamma(c_j)} \frac{d_j^{c_j}}{\left(d_j + \sum_{i=1}^{m} f_i^{\beta_j} + \sum_{u=1}^{l} s_u^{\beta_j} \right)^{m+c_j}}} \tag{8}$$

The conditional posterior of λ is $p(\lambda, \beta_j / f^{(m)}, s^{(l)}, r^{(n)}) / p(\beta_j / f^{(m)}, s^{(l)}, r^{(n)})$,

$$p(\lambda / \beta_j, f^{(m)}, s^{(l)}, r^{(n)}) =$$

$$\frac{\left(d_j + \sum_{i=1}^{m} f_i^{\beta_j} + \sum_{u=1}^{l} s_u^{\beta_j} \right)^{m+c_j}}{\Gamma(m+c_j)} \lambda^{m+c_j-1} \exp\left[-\lambda \left(d_j \sum_{i=1}^{m} f_i^{\beta_j} + \sum_{u=1}^{l} s_u^{\beta_j} \right) \right] \tag{9}$$

so that

$$(\lambda / \beta_j, f^{(m)}, s^{(l)}, r^{(n)}) \sim \text{Gamma}\left(m + c_j, d_j + \sum_{i=1}^{m} f_i^{\beta_j} + \sum_{u=1}^{l} s_u^{\beta_j} \right).$$

3.4 The Posterior Distribution for Interval Censored Data

For interval censored data $m^{(i-1)} = (m_1, m_2, \ldots, m_{i-1})$, where m_j is the observed number of failures in $[T_{j-1}, T_j)$, $j = 1, 2, \ldots, i - 1$, Aboura [2] determines the posterior distribution of (λ, β) as

$$p(\lambda, \beta / m^{(i-1)}) \propto \prod_{j=1}^{i-1} \binom{M}{m_j} (1 - e^{-\lambda \Delta T_j^{\beta}})^{m_j} (e^{-\lambda \Delta T_j^{\beta}})^{M-m_j} g(\lambda, \beta / r^{(n)}) \tag{10}$$

M being the total number of items and $\Delta T_j = T_j - T_{j-1}$. For small i and M, a closed form expression can be obtained for the posterior distribution through an extension of (10) in a Binomial sum. If i and M are realistically large, the number of summations in the closed form expression would become large and not practical to carry. A two-dimensional numerical computation provides $p(\lambda, \beta / m^{(i-1)})$.

4 The Replacement Strategy

The maintenance optimization procedure consists of determining at each planned replacement time T_{i-1}, the next preventive replacement time T_i,

$i = 1, 2, \ldots$. In between the prescribed times $T_1, T_2, \ldots$, replacement or repair are not made upon failure of the operating item. In the traditional maintenance optimization approach, the time intervals between planned replacements would be equal and determined at the start of the operations. In the approach presented here, the replacement times are determined a stage ahead. The adaptive nature of the policy reduces the economic loss of a fixed time replacement protocol, as the maintenance procedure reaches for an optimal replacement time. The expected cost per unit of time for the next replacement interval is minimized, the expectation taken with respect to past failure and survival information.

4.1 Cost Function

At time T_{i-1}, $i = 1, 2, \ldots$, the expected cost per unit of time in the interval $[T_{i-1}, T_i)$, is

$$c(\Delta T_i) = \frac{c_p + c_f E(N_i/D_{i-1})}{\Delta T_i} \tag{11}$$

where c_p is the cost of replacing all M items, c_f is the cost of failure of one item and $E(N_i/D_{i-1})$ is the expected number of failures in $[T_{i-1}, T_i)$, the expectation taken with respect to the available information D_{i-1}. D_0 is the set of all relevant information known prior to and at time T_0. $D_0 = \{r^{(n)}, t^{(n)}, b\}$. In the case of complete data, $D_{i-1} = \{f^{(m)}, s^{(l)}\} \cup D_0$. $f^{(m)} = (f_1, f_2, \ldots, f_m)$ and $s^{(l)} = (s_1, s_2, \ldots, s_l)$ are the failure and survival times respectively, observed in $(T_0, T_{i-1}]$. In the case of censored data, $D_{i-1} = \{m^{(i-1)}\} \cup D_0$, where $m^{(i-1)} = (m_1, m_2, \ldots, m_{i-1})$, m_j being the observed number of failures in $[T_{j-1}, T_j)$, $j = 1, 2, \ldots, i-1$. The replacement cost c_p is relatively easy to assess. c_p consists of the cost of the M new items and the cost of manhours needed for the replacement. c_f is the cost due to the failure of one item. Although such a cost often exists in the form of inconvenience, reduced productivity or other, it may be hard to attach a dollar value to it. In Section 5, we consider several cost models.

4.2 Maintenance Constraint

If c_p and c_f are constants in (11), and since N_i, $i = 1, 2, \ldots$, the number of failed items in $[T_{i-1}, T_i)$, is bounded by M, $c(\Delta T_i)$ approaches 0 as ΔT_i becomes large. In practice, it makes sense to adopt the present maintenance scenario only if an acceptable proportion of items is to fail in $[T_{i-1}, T_i)$. To model this constraint, c_f can be made to increase with the number of failed items N_i. This penalty function may be hard to assess. Instead, a constraint on the number of failures N_i can be introduced. For example, the probability of the number of failures N_i not exceeding a predetermined number K may be constrained to be at least $1 - \alpha$, $0 < \alpha < 1$. The problem parameters K and α would be determined by the maintenance management. Another possible constraint on the number of failures N_i, is to restrain the expected value $E(N_i/D_{i-1})$ to be less than some number K. These constraints would force ΔT_i to be bounded above, rendering the minimization of $c(\Delta T_i)$ possible. In the next section, we consider four cases where (i) the cost of failure c_f is zero and the expected number of failures is constrained, (ii) the cost of failure c_f is a constant and the expected number of failures is constrained, (iii) the cost of failure c_f is a penalty function for the number of failures, and (iv) a penalty cost occurs if the number of failures exceeds a predetermined number.

5 Maintenance Optimization Models

5.1 The Constant Cost Model

The Constant Cost model considers the case in which there is no dollar value associated with the failure of an item ($c_f = 0$). The range of possible time intervals before the next replacement is limited by a constraint on the expected number of failures in the considered interval. The maintenance optimization problem at time T_{i-1} is

$$\min_{\Delta T_i > 0} c_p / \Delta T_i \text{ subject to } E(N_i/D_{i-1}) \leq K \tag{12}$$

In both data collection cases $E(N_i/D_{i-1}) \leq K$ translates into $R(\Delta T_i/D_{i-1}) \geq 1 - K/M$, where $R(t/D_{i-1})$ is the reliability for mission time t as assessed at time T_{i-1}. Since

$$E(N_i/D_{i-1}) = \sum_{j=1}^{k} \int_0^{\infty} E(N_i/\lambda, \beta_j) g(\lambda, \beta_j/D_{i-1}) d\lambda, \quad j = 1, 2, \ldots, k,$$

$$E(N_i/\lambda, \beta_j) = M(1 - e^{-\lambda \Delta T_i^{\beta}}) = M(1 - R(\Delta T_i/\lambda, \beta_j))$$

and

$$E(N_i/D_{i-1}) = M(1 - R(\Delta T_i/D_{i-1}))$$

Therefore the optimal time interval is the largest feasible ΔT_i value. The optimal time intervals ΔT_i, $i = 1, 2, \ldots$, depend only on the ratio K/M

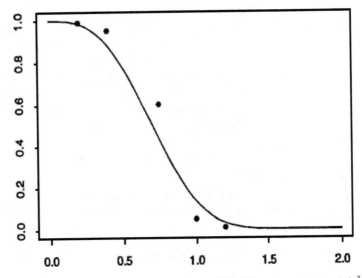

Figure 1 The initial reliability estimates $(t^{(5)}, r^{(5)})$ plotted against e^{-2t^3}.

and the reliability function $R(\cdot/D_{i-1})$. If λ and β were to be known, then $R(\Delta T_i/D_{i-1}) = R(\Delta T_i/D_\infty) = e^{-\lambda \Delta T_i^\beta}$, $i = 1, 2, \ldots$, and $\Delta T_\infty = (-\ln(1-K/M)/\lambda)^{1/\beta}$ would be the optimal time interval between replacements, at all stages. ΔT_∞ is the value to which ΔT_i ultimately converges to in both data collection, given the assumption of a Weibull lifetime model.

5.1.1 Simulation of the Maintenance Optimization Procedure

In a simulated example where $\lambda = 2$, $\beta = 3$ and $K/M = 0.2$, we assume that reliability estimates $r^{(5)} = (0.99, 0.95, 0.60, 0.05, 0.01)$ for mission times $t^{(5)} = (0.2, 0.4, 0.75, 1, 1.2)$ are given by an expert or taken from some other knowledgeable source. Figure 1 shows the expert input $(t^{(5)}, r^{(5)})$ plotted against the reliability function e^{-2t^3}. The resulting prior distribution $g(\lambda, \beta/r^{(n)})$ [2] is shown in Figure 2.

The optimal value for the first replacement time T_1 is obtained as the time t at which the prior reliability $R(t/D_0)$ is equal to $1 - K/M = 0.8$. Therefore in this example the first replacement of all items is to occur at time $T_1 = \Delta T_1 = 0.265$. The prior reliability function $R(t/D_0)$ is shown in Figure 3 with the resulting optimal first time interval $\Delta T_1 = 0.265$. The

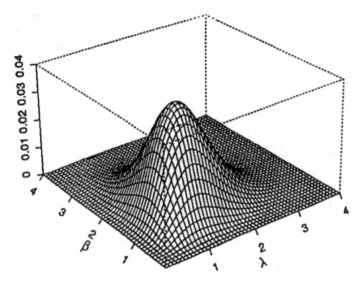

Figure 2 The prior distribution $g(\lambda, \beta/r^{(n)})$.

dashed line function in the graph of Figure 3 is e^{-2t^3} with the corresponding limiting optimal time interval $\Delta T_\infty = (-\ln(1 - 0.2)/2)^{1/3} = 0.481$.

At the successive times T_{i-1}, $i = 2, 3, \ldots$, the optimal time intervals ΔT_i are obtained as the solution of $R(\Delta T_i/D_{i-1}) = 1 - K/M$. In the case of complete data $R(\Delta T_i/D_{i-1})$ obtains in a closed algebraic form while it must be numerically evaluated at each stage, in the case of interval censored data [2].

Complete Data

$$R(\Delta T_i/D_0) = \sum_{j=1}^{k} \int_0^\infty e^{-\lambda \Delta T_i^\beta}) g(\lambda, \beta_j/D_{i-1}) d\lambda$$

$$= \sum_{j=1}^{k} \left(\int_0^\infty e^{-\lambda \Delta T_i^\beta} \frac{d_j^{c_j} \lambda^{c_j-1} e^{-d_j\lambda}}{\Gamma(c_j)} d\lambda \right) q_j$$

$$= \sum_{j=1}^{k} \frac{d_j^{c_j} q_j}{(d_j + \Delta T_i^{\beta_j})^{c_j}}$$

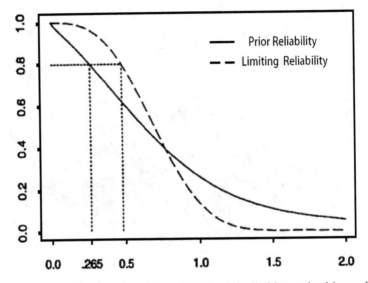

Figure 3 The first replacement time $\Delta T_1 = 0.265$ and the limiting optimal interval $\Delta T_\infty = 0.481$.

$$
\times \left(\int_0^\infty \frac{(d_j + \Delta T_i^{\beta_j})^{c_j} \lambda^{c_j - 1} e^{-\lambda(d_j + \Delta T_i^{\beta_j})}}{\Gamma(c_j)} d\lambda \right)
$$

$$
= \sum_{j=1}^k \left(\frac{d_j}{d_j + \Delta T_i^{\beta_j}} \right)^{c_j} q_j
$$

For $i = 2, 3, \ldots, R(\Delta T_i / D_{i-1}) = R(\Delta T_i / D_0, f^{(m)}, s^{(l)})$, and

$R(\Delta T_i / D_0, f^{(m)}, s^{(l)})$

$$
= \sum_{j=1}^k \left(\int_0^\infty e^{-\lambda \Delta T_i^\beta} p(\lambda / \beta_j, f^{(m)}, s^{(l)}) d\lambda \right) p(\beta_j / f^{(m)}, s^{(l)})
$$

$$
= \sum_{j=1}^k \left(\int_0^\infty e^{-\lambda \Delta T_i^\beta} \frac{(d_j + \sum_{i=1}^m f_i^{\beta_j} + \sum_{u=1}^l s_u^{\beta_j})^{m+c_j}}{\Gamma(m + c_j)} \right.
$$

$$
\left. \times \lambda^{m+c_j-1} \exp\left[-\lambda \left(d_j + \sum_{i=1}^m f_i^{\beta_j} + \sum_{u=1}^l s_u^{\beta_j} \right) \right] d\lambda \right)
$$

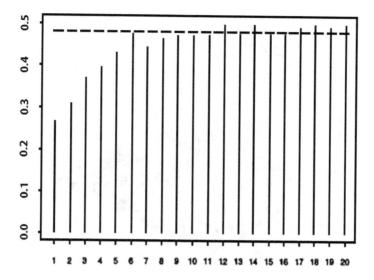

Figure 4 The optimal time intervals ΔT_i, $i = 1, 2, \ldots, 20$ for $M = 10$, $K = 2$ and $c_p = 30$.

$$p(\beta_j / f^{(m)}, s^{(l)})$$

$$= \sum_{j=1}^{k} \frac{\left(d_j + \sum_{i=1}^{m} f_i^{\beta_j} + \sum_{u=1}^{l} s_u^{\beta_j}\right)^{m+c_j}}{\left(d_j + \sum_{i=1}^{m} f_j^{\beta_j} + \sum_{u=1}^{l} s_u^{\beta_j} + \Delta T_i^{\beta}\right)^{m+c_j}} p(\beta_j / f^{(m)}, s^{(l)})$$

The optimal time intervals between replacements, ΔT_i, $i = 1, 2, \ldots, 20$, are plotted in Figure 4 for a 20 stages simulation of the maintenance routine. In this example $M = 10$, $K = 2$ and $c_p = 30$. The exact failure times are recorded between the replacement times. The horizontal dashed line in Figure 4 marks the limiting optimal time interval $\Delta T_\infty = 481$. As data is gathered between the replacement times, the optimal time intervals improve to finally stabilize around the limiting value. Figure 5 shows the prior distribution and the posterior distribution of (λ, β) at time T_{19}.

Figures 6 and 7 show the optimal time intervals ΔT_i, $i = 1, 2, \ldots, 10$, for $M = 40$, $K = 8$, and $M = 100$, $K = 20$, respectively.

The adaptive nature of the maintenance procedure is observed in the two examples. Similar behavior of the maintenance optimization procedure was observed in the case of interval censored data collection.

To study the convergence of the adaptive procedure, 100 simulation runs were made for both data collection cases. Each replication consisted

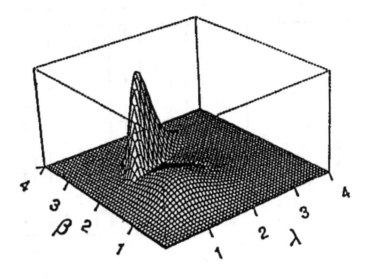

Figure 5 Prior and posterior distribution of (λ, β) at time T_{19}.

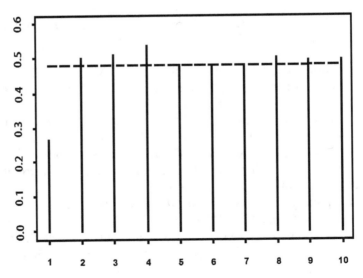

Figure 6 The optimal time intervals ΔT_i, $i = 1, 2, \ldots, 20$ for $M = 40$, $K = 8$.

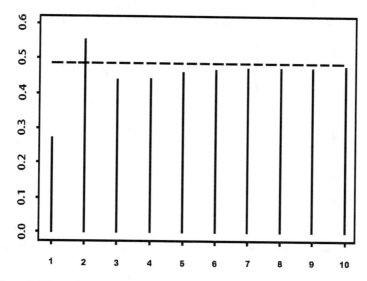

Figure 7 The optimal time intervals ΔT_i, $i = 1, 2, \ldots, 10$ for $M = 100$, $K = 2$.

of the simulation of 10 replacement stages for $M = 100$, $K = 20$. Figure 8 shows the maximum, mean and minimum optimal time interval ΔT_i, $i = 1, 2, \ldots, 10$ for the case of complete data observation, the statistics computed over the 100 replications. Figure 9 shows the equivalent of Figure 8 for the interval censored data collection scenario. In both cases, the maintenance optimization procedure converges rapidly in the mean value, with the spread around the mean values getting smaller as more information is gathered through the replacement stages. The case of interval censored data shows more spread around the mean values, as expected.

Note that although the expert reliability estimates indicate an optimistic opinion for the early life of the item, the first optimal replacement time $\Delta T_1 = 0.265$ is a cautious decision, being smaller than 0.481. This is due to the fact that the prior variance parameter b has been set to 0.001. This choice of value for b gives a prior mean of 1.506 for λ and a prior mean of 1.576 for β with relatively large prior variances. If b was made very large, a prior distribution degenerate at (2.926, 5.831) would have resulted for (λ, β), with a first replacement time equal to 0.643. The smaller b, the more variance is allowed in the prior distribution providing a better adaptation to observed data. Table 1 shows the prior means and variances of λ and β for different values of b.

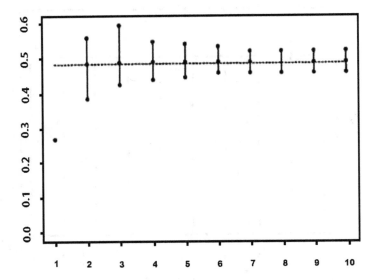

Figure 8 The maximum, mean and minimum optimal time interval ΔT_i, $i = 1, 2, \ldots, 10$ for the case of complete data observation.

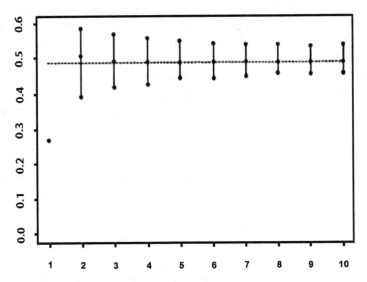

Figure 9 The maximum, mean and minimum optimal time interval ΔT_i, $i = 1, 2, \ldots, 10$ for the case of the interval censored data.

Table 1 Prior means and variances of λ and β for different values of the parameter b.

b	10^{-10}	10^{-3}	1	10	100	10^3	10^4	2×10^4
λ Mean	1.506	1.506	1.615	2.002	2.478	2.854	2.923	2.926
λ Variance	0.377	0.377	0.347	0.198	0.070	0.015	0.001	8.26×10^{-4}
β Mean	1.576	1.576	1.854	3.067	4.677	5.660	5.823	5.831
β Variance	0.372	0.372	0.396	0.362	0.236	0.054	0.006	2.82×10^{-3}

5.2 The Constant Failure Cost Model

The Constant Failure Cost model considers the case in which a constant cost c_f is associated with the failure of each item. The range of possible time intervals before the next replacement remains limited by a constraint on the expected number of failures in the considered interval. The maintenance optimization problem at time T_{i-1} is

$$\min_{\Delta T_i > 0} \frac{c_p + c_f E(N_i / D_{i-1})}{\Delta T_i}$$

$$\text{subject to } E(N_i / D_{i-1}) \le K \tag{13}$$

Figure 10 shows the first optimal interval $\Delta T_1 = 0.265$ and the optimal time interval given knowledge of $\lambda = 2$, $\beta = 3$, $\Delta T_\infty = 0.361$ for the example of Section 5.1 when $c_p = 30$ and $c_f = 18$. The full line in Figure 10 is the cost function of (13) for $i = 1$. The dashed line is the limiting cost function $(c_p + c_f M(1 - R(t/D_\infty)))/t = (c_p + c_f M(1 - e^{-2t^3}))/t$. Given the positive cost of failure in this model, the limiting optimal time interval $\Delta T_\infty = 0.361$ is less than its value 0.481 obtained in the case of the Constant Cost model. As failure information is gathered, the cost function approaches its limit shown by the dashed line in Figure 10, making the optimal time interval ΔT_i converge to its limiting value $\Delta T_\infty = 0.361$.

5.3 The Penalty Cost Model I

This penalty cost model assumes an increasing cost function $c_f(j)$ for the failure of j items. Various functions may be chosen to model $c_f(j)$ depending on the application. The maintenance optimization problem at time at time T_{i-1} is

$$\min_{\Delta T_i > 0} \frac{c_p + \sum_{j=0}^{M} c_f(j) \operatorname{Prob}(N_i = j/D_{i-1})}{\Delta T_i} \tag{14}$$

where

$$\operatorname{Prob}(N_i = j/D_{i-1})$$

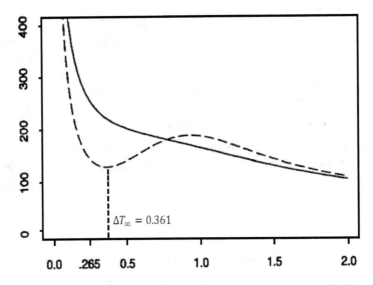

Figure 10 The first optimal interval $\Delta T_1 = 0.265$ and the optimal interval $\Delta T_\infty = 0.361$.

$$= \sum_\beta \int_\lambda \binom{M}{j} (1 - e^{-\lambda \Delta T_i^\beta})^j (e^{-\lambda \Delta T_i^\beta})^{M-j} g(\lambda, \beta/r^{(n)})d\lambda \quad (15)$$

Although (14) is an unconstrained minimization problem, it can be seen that for an appropriate choice of the penalty function c_f, an optimal solution ΔT_i obtains for each stage $i = 1, 2, \ldots$, etc. The determination of the penalty function $c_f(\cdot)$ may not be a trivial task and therefore makes this model hard to apply.

5.4 The Penalty Cost Model II

A more applicable penalty cost model is a special case of that of Section 5.3, in which a penalty is applied when the number of failures exceeds a predetermined number. An example is when a city council imposes such a penalty on the electricity company for not keeping a certain percentage of the city lights operating in a particular area of the city. In this case, the maintenance optimization problem at time T_{i-1} is

$$\min_{\Delta T_i > 0} \frac{c_p + c_f \, \text{Prob}(N_i > K/D_{i-1})}{\Delta T_i} \quad (16)$$

For (16) to have a solution, a constraint needs to be imposed, say $\Delta T_i \leq \delta$ for some δ to be defined by the maintenance management.

6 Concluding Remarks

An adaptive procedure for the optimal replacement of identical items operating under similar conditions was outlined. Four maintenance optimization models were developed. Two data collection scenarios were considered. The procedure was demonstrated and its convergence shown in both data collection cases in simulated examples. The procedure is easy to implement and can result in substantial savings. The adaptive nature of the procedure is a modern feature that permits an updating of the lengths of times between replacements as failure information is gathered. The methodology in this paper was developed following technical discussions with an electricity company.

References

[1] K. Aboura. Use of informed judgement in the assessment of a prior distribution for the parameters of the Weibull and eneralized Gamma Reliability Models. CSIRO Report No. DMS-D 95/63, 1995.

[2] K. Aboura. Bayesian adaptive maintenance plans using initial expert reliability estimates. CSIRO Report No. DMS-D 95/64, 1995.

[3] G. Aras, L. R. Whitaker, and Y.-H. Wu. Sequential nonparametric estimation of an optimal age replacement policy: A simulation study. Communications in Statistics – Simulation and Computation, 22(4):1115–1134, 1993.

[4] W. M. Bassin. A Bayesian optimal overhaul interval model for the Weibull restoration process case. Journal of the American Statistical Association, 68(343):575–578, 1973.

[5] J. A. Bather. On the sequential construction of an optimal age replacement policy. Bulletin of Institute of International Statistics, 47:253–256, 1977.

[6] A. Benmerzouga. Optimal group replacement policies. PhD Dissertation, Case Western Reserve University, 1991.

[7] S. Dayanik and U. Gurler. An adaptive Bayesian replacement policy with minimal repair. Operations Research, 50(3):552–558, 2002.

[8] R. Flage, D. W. Coit, J. T. Luxhøj, and T. Aven. Safety constraints applied to an adaptive Bayesian condition-based maintenance optimization model. Reliability Engineering and System Safety, 102:16–26, 2012.

[9] B. Fox. Adaptive age replacement. Journal of Mathematical Analysis and Applications, 18:365–376, 1967.

[10] E. W. Frees and D. Ruppert. Sequential nonparametric age replacement policies. The Annals of Statistics, 13(2):650–652, 1985.

[11] M.-G. Juang and G. Anderson. A Bayesian method on adaptive preventive maintenance problem. European Journal of Operational Research, 155:455–473, 2004.

[12] D. V. Lindley. Reconciliation of probability distributions. Operations Research, 31:866–880, 1983.

[13] T. A. Mazzuchi and R. Soyer. A Bayesian perspective on some replacement strategies. Reliability Engineering and System Safety, 51:295–303, 1996.

[14] A. Sarkar, S. C. Panja, and B. Sarkar. Survey of maintenance policies for the last 50 years. International Journal of Software Engineering & Applications, 2(3):139–148, 2011.

[15] S.-H. Sheu, R. H. Yeh, Y.-B. Lin, and M.-G. Juan. A Bayesian perspective on age replacement with minimal repair. Reliability Engineering and System Safety, 65:55–64, 1999.

[16] S.-H. Sheu, R. H. Yeh, Y.-B. Lin, and M.-G. Juan. A Bayesian approach to an adaptive preventive maintenance model. Reliability Engineering and System Safety, 71:33–44, 2001.

[17] R. M. Soland. Bayesian analysis of the Weibull process with unknown scale and shape parameters. IEEE Transactions on Reliability, 18:181–184, 1969.

[18] C. Valdez-Flores and R. M. Feldman. A survey of preventive maintenance models for stochastically deteriorating single-unit systems. Naval Research Logistics, 36(4):419–446, 1989.

[19] M. West. Modelling expert opinion. In J. M. Bernardo, M. H. Degroot, D. V. Lindley, and A. F. M. Smith (Eds.), Bayesian Statistics, Vol. 3, pp. 493–508. Oxford University Press, 1988.

[20] J. G. Wilson and E. Popova. Bayesian approaches to maintenance intervention. In Proceedings of the Section on Bayesian Science of the American Statistical Association, pp. 278–284, 1998.

[21] E. Zio and M. Compare. Evaluating maintenance policies by quantitative modeling and analysis. Reliability Engineering and System Safety, 109:53–65, 2013.

Biographies

Khalid Aboura teaches quantitative methods at the College of Business Administration, University of Dammam, Kingdom of Saudi Arabia. Dr Khalid Aboura spent several years involved in academic research and consulting at the George Washington University, Washington DC, USA, where he completed the Master of Science and the Doctor of Science degrees in Operations Research. Dr. Aboura has extensive experience in Stochastic Modelling in Operations Research and Engineering, Simulation, Maintenance Optimization and Mathematical Optimization. He served as Chairman of the Statistical Computing Section of the Washington Statistical Society. Khalid Aboura worked as a Research Scientist at the Division of Mathematics and Statistics of the Commonwealth Scientific and Industrial Research Organization (CSIRO) of Australia. During his tenure at CSIRO, Khalid Aboura was involved in research and consulting with the Australian industry on a number of projects. Khalid Aboura also conducted research at the School of Civil and Environmental Engineering and at the School of Computing and Communication, Faculty of Engineering and Information Technology, University of Technology Sydney, Australia. In China, Khalid Aboura was a Scientist at the Kuang-chi Institute of Advanced Technology of Shenzhen. His involvement was in the development of Computer Model Emulation solutions for the rapid design of metamaterials.

Johnson I. Agbinya is currently Associate Professor in the department of electronic engineering at La Trobe University, Melbourne Australia. He is also Honorary Professor at the University of Witwatersrand (WITS), South Africa; Extraordinary Professor at the University of the Western Cape (UWC), Cape Town and the Tshwane University of Technology (TUT), Pretoria, South Africa. Prior to joining La Trobe University in November 2011, he was Senior Research Scientist at CSIRO Telecommunications and Industrial Physics (now CSIRO ICT) from 1993–2000, Principal Research Engineering at Vodafone Australia (2000–2003) and Senior Lecturer at UTS Australia (2003–2011). His R&D activities cover remote sensing, Internet of things (machine to machine communications), bio-monitoring systems, wireless power transfer, mobile communications and biometrics systems. He has authored/co-author nine technical books in telecommunications, some of which are used as textbooks. He is founder of the International conference on broadband communications and biomedical applications (IB2COM), Pan African conference on science, computing and telecommunications (PACT) and the African Journal of Information and Communication Technology (AJICT). He has published more than 250 peer-reviewed research publications in international Journals and conference papers. He received his BSc degree electronic/electrical engineering from Obafemi Awolowo University (OAU), Ile Ife, Nigeria; MSc in electronic engineering from the University of Strathclyde, Glasgow Scotland and PhD from La Trobe University, Melbourne, Australia in 1973, 1982 and 1994 respectively. He received Best Paper award from IEEE 5th International Conference on Networking (ICN' 2006) Mauritius, CSIRO ADCOM group research award in 1997 and Research Trailblazer Certificate at UTS in 2009. He is the Editor in Chief of the African Journal of ICT (AJICT), General Chair of several international conferences and member of several current international technical conference committees. He has served as expert on several international grants reviews/committees and was a rated researcher by the South African National Research Fund (NRF).

Heart Rate Variability, Blood Pressure and Cognitive Function: Assessing Age Effects

Louisa Giblin, Levin De Leon, Lisa Smith, Tamara Sztynda
and Sara Lal

*School of Medical and Molecular Biosciences, University of Technology Sydney,
P.O. Box 123, Broadway, NSW 2007, Australia; e-mail: sara.lal@uts.edu.au*

Received 15 February 2013; Accepted 27 March 2013

Abstract

Increasing age is the most significant risk factor for dementia. Aging populations see cognitive disorders becoming increasingly prevalent, unfortunately paired with high economic and social consequences. Mild cognitive impairment (MCI) is the earliest detectable stage preceding dementia. This study aimed to identify early links between heart rate variability (HRV) and blood pressure (BP) to cognitive performance. Three blood pressure readings were taken pre and post study. Electrocardiogram was recorded during both resting (baseline) and cognitive interventions (active). HRV was extrapolated using a fast Fourier transform algorithm to produce low and high frequency bandwidths. Two psychometric tools were administered to assess cognitive domains such as memory, reasoning and visual construction ability. In the youngest age group, 18–35 years, higher blood pressure was detrimental to judgment and orientation but beneficial to calculation and memory skills. Higher sympathetic drive (low frequency) impaired language, recall and attention ability. In the middle age group (36–50 years) higher blood pressure predicted decline in comprehension, orientation and attention domains. Higher sympathetic activity (low frequency) was linked to decreases in various domains such as similarity and construction. The oldest group (51–65 years) showed higher blood pressure precipitated declines in recall ability and high sympathetic activity (low frequency) impaired orientation func-

Journal of Green Engineering, Vol. 3, 347–361.

tion. These various associations suggest autonomic activity biomarkers for cognitive impairment vary according to age. Few studies confirm specific autonomic implications on cognition from young to older age. The cognitive associations reported highlight the potential importance of autonomic activity as a predictive tool for cognitive decline. Early detection of cognitive impairment allows for intervention methods to be applied sooner to slow or cease cognitive decline progression.

Keywords: Aging, cognition, heart rate variability.

1 Introduction

Although aging naturally involves a degree of cognitive decline, mild to severe cognitive impairment is not considered a healthy progression of aging. Increasing age is the most significant risk factor for dementia. As global populations age, dementia prevalence is predicted to increase four-fold to 115 million people by 2050 [1]. As Alzheimer's disease (the most common form of dementia) has no known cure, early detection and prevention is crucial.

Mild cognitive impairment (MCI) is the earliest detectable stage preceding dementia onset. MCI may impede (not prevent) daily life functioning and may manifest by symptoms of memory loss and subtle difficulties performing complex cognitive tasks in domains such as attention and language. Transition rates of MCI into dementia have shown that almost 50% of MCI patients develop dementia within five years [17]. This suggests that persons with MCI may also have preclinical dementia. Early detection of this vulnerable stage would allow for earlier application of preventative treatments. The present study investigated heart rate variability (HRV) analysis as a novel physiological marker to identify those at higher risk of MCI.

In contrast to MCI, dementia impairs daily life functioning. It has extensive effects, both socially and economically [25]. Many risk factors have been established which contribute to Alzheimer's type dementia, the most prominent being increasing age. Other risk factors include female sex; presence of apolipoprotein E ε4 allele; low education level and cardiovascular disease [5].

The aging process involves a gradual 'undoing' of the body, resulting in various chemical and physical alterations which contribute to cognitive decline, although there remains a difference between natural cognitive decline and cognitive impairment [8]. In an autonomic sense, aging is accompanied by a degree of parasympathetic withdrawal [32] which can lead to hypertension due to resultant sympathetic dominance. High BP is hypothesised to contribute to cognitive impairment through subtle disturbances in cerebral perfusion, thereby altering the neuron's biochemical environment and optimal functioning. Neuroimaging studies of hypertensive patients have shown decreased cerebral oxygen metabolism, enlarged ventricles (cerebral atrophy) and increased white matter lesions [45]. These factors often occur unbeknownst to the individual until more severe cognitive symptoms develop (potentially years later), as a cumulative manifestation of neuroanatomical changes.

There remains contention in the area as many studies report variations in the relationships between BP and cognition. Most studies suggest high BP is a major risk factor (especially midlife) [4, 47], whereas others suggest low BP is more detrimental (particularly in older age) [19, 33]. Waldstein et al. [51] found links between both high and low BP impairing cognitive function (inverted U-shaped hypothesis) or no associations existing at all [10]. Most studies reinforce the importance of early BP control to reduce the development of cognitive symptoms in later life.

Heart rate variability is a physiological measurement reflecting the autonomic balance of the heart [38]. HRV is derived by the spectral analysis of millisecond time variations between consecutive heart beats. These variations reflect the interplay between the parasympathetic (high frequency (HF)) and sympathetic (low frequency (LF)) branches. The sinus node of the heart acts as a pacemaker, regulating contractions to accommodate metabolic demand. It is densely innervated by both the parasympathetic and sympathetic divisions of the ANS [38]. Sympathetic innervation from the stellate ganglia is mediated by noradrenalin release at the sinus node, which is metabolised relatively slowly, as opposed to parasympathetic activation of the heart, which is moderated by the vagus nerve via acetylcholine release and quickly metabolised [38]. The distinct turnover rates of the two chemical transmitters result in variations between frequencies and fluctuations of heart rate producing a complex variability characterised by HRV analysis. These variations have been identified and quantified to establish different bandwidth frequency standards at which the two autonomic subsystems function [48]. Baseline heart rate is driven by parasympathetic activity, known as tonic inhibitory control.

Clinical applications of HRV include detection of autonomic neurodegeneration in diabetic patients [15] and clinical risk assessments of cardiac related mortalities [49]. This highlights the application of HRV as a predictive clinical tool. Studies show that changes in autonomic activity in early adulthood and midlife increase the risk of cognitive impairment developing later in life. In particular, low HRV has been linked with poor cognition, where the autonomic system is less reactive to changes in the external environment and is therefore less adaptable [37]. Low HRV has also been proposed as a marker of disease in many studies [11, 14, 23, 50]. Few studies, however, assess autonomic nervous system activities as predictive risk factors for the development of cognitive impairment. The present study aimed to address this gap in the literature by identifying the relationships between cognitive function and cardiac autonomic activity (HRV and BP) over a range of ages: young (18–35 years), middle (36–50 years) and older age (51–65 years). A focus on detection of the early stages of MCI is integral, before progression to a less treatable and functional state.

2 Methods

A total of 51 participants were recruited from the community adding to an existing database of 100 [9, 46] to produce a cumulative total of $n = 151$. Firstly three blood pressure (BP) measurements were taken (pre-study average) followed by a three-lead electrocardiogram. The participant underwent two interventions during which electrocardiogram was measured with eyes open; resting (baseline) state and during a cognitive task (active neutral conversation [34]). The electrocardiogram data was used to extrapolate HRV by spectral analysis of the time variations between consecutive R waves of the QRS complex. The electrocardiogram is first applied with the Butterworths filter, a band pass filter to diminish frequencies below 2 Hz and above 40 Hz, to reduce movement artefacts, T-wave interference and electrocardiogram baseline drift from influencing the data [36]. The data is then applied with a non-linear squaring function in preparation for application of the fast Fourier transform (non-parametric) [36]. The fast Fourier transform produces a spectrogram that models the power densities of the R-R intervals from the electrocardiogram (Figure 1).

Heart rate variability data reflects sympathetic (LF: 0.04–0.15 Hz) and parasympathetic (HF: 0.15–0.5 Hz) branches of the autonomic nervous system [27]. Sympathovagal balance, a measure of the sympathetic and parasympathetic equilibrium, was also determined (LF:HF) [12]. Total power (TP) reflects the total area under the spectrogram curve.

Cognitive function was assessed using two psychometric evaluation tools: the Mini-Mental State Examination [16] and the Cognistat [13]. These validated and reliable cognitive tests were administered in conjunction with one

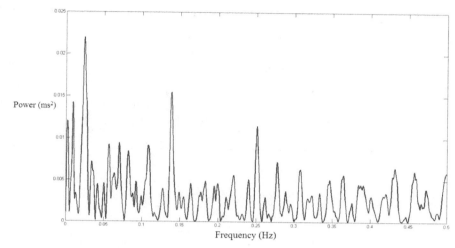

Figure 1 Heart rate variability power spectrogram of a participant during the cognitive task (active).

another to increase sensitivity and specificity of the cognitive outcomes [43]. The tests assess cognitive function in domains such as memory, language, judgement and calculation. A final three BP readings were taken (post-study average), completing the study protocol. Differences between age groups were assessed by Analysis of Variance with post-hoc least significant difference Fisher analysis. Relationships between cognitive function and other variables (BP and HRV) were assessed using Pearson's correlations. Where a cognitive domain was significantly linked to three or more other variables, a regression analysis was performed to determine the most significant predictor. All significant findings were reported at p values of < 0.05.

3 Results

Blood pressure increased significantly with increasing age yet affected cognitive domains differently over the life span. Results suggest aging and age-related BP levels are significant predictors for cognitive impairment, with performance in domains of language ($p = 0.0002$), orientation ($p = 0.002$), construction ($p = 0.03$) and total score ($p = 0.01$) declining with aging. Significant findings between BP and cognitive function for the three age groups are shown in Table 1. Heart rate variability effects on cognition during baseline were assessed among the three age groups, significant findings are

Table 1 Comparison between blood pressure and cognitive domain performance in three age groups.

Age group	Blood pressure	Cognitive domain	r	p
18-35 years	Systolic	Memory	0.3	0.02
		Recall	0.29	0.03
		Repetition	0.39	0.002
	Diastolic	Calculation	0.25	0.048
		Orientation	-0.32	0.01
		Judgment	-0.3	0.02
36-50 years	Systolic	Comprehension	-0.37	0.01
	Diastolic	Similarity	0.31	0.03
		Comprehension	-0.29	0.05
		Orientation	-0.35	0.02
		Attention	-0.29	0.04
		Calculation	0.28	0.049
51-65 years	Systolic	Recall	-0.39	0.01
	Diastolic	Recall	-0.34	0.03

presented in Table 2. A regression analysis performed found 13% of the variance of the total MMSE scores (36–50 years) were predicted by LF, HF and TP collectively ($F = 1.95$, $df = (3, 39)$, $p < 0.14$, $R = 0.36$, $R^2 = 0.13$, adjusted $R^2 = 0.64$). The multiple regression for the similarity domain for 36–50 years had an overall significance of $p < 0.004$. The regression identified significance for three HRV variables (LF, HF and TP) which together explained 28% of the variance in the similarity domain ($F = 5.14$, $df = (3, 39)$, $p < 0.004$, $R = 0.53$, $R^2 = 0.28$, adjusted $R^2 = 0.23$). However, individually two HRV factors, LF ($p < 0.02$) and TP ($p < 0.02$) were the strongest significant predictors The 36–50 year group also showed 10% of the variability of the total Cognistat score was explained by LF, HF and TP ($F = 1.2$, $df = (3, 31)$, $p < 0.33$, $R = 0.32$, $R^2 = 0.1$, adjusted $R^2 = 0.02$). Cognitive performance and HRV associations were also identified during the active cognitive task (Table 3).

Table 2 Comparison between heart rate variability (baseline) and cognitive domain performance in three age groups.

Age group	Baseline HRV parameter	Cognitive domain	r	p
18-35 years	HF	Construction	-0.36	0.01
		Naming	-0.45	0.001
	LF	Recall	-0.4	0.003
	TP	Construction	-0.3	0.03
		Recall	-0.32	0.02
36-50 years	HF	Judgment	-0.32	0.046
		Similarity	-0.43	0.01
		Total score (Cognistat)	-0.34	0.03
		Total score (MMSE)	-0.33	0.04
	LF	Attention	-0.38	0.02
		Construction	-0.33	0.04
		Repetition	-0.36	0.02
		Similarity	-0.32	0.04
		Total score (Cognistat)	-0.32	0.04
		Total score (MMSE)	-0.37	0.02
	TP	Attention	-0.39	0.01
		Construction	-0.34	0.03
		Judgement	-0.32	0.046
		Repetition	-0.41	0.01
		Repetition	-0.41	0.01
		Similarity	-0.44	0.01
		Total score (Cognistat)	-0.41	0.01
		Total score (MMSE)	-0.39	0.01
51-65	LF	Orientation	-0.53	0.001

Key: HF=High frequency; LF=Low frequency; MMSE=Mini Mental State Examination; TP=Total power

4 Discussion

4.1 Blood Pressure and Cognitive Function

Many studies report that autonomic cardiovascular factors may affect cognitive function, particularly BP levels [44, 52]. The majority of research is confined to assessing BP effects on cognition on those aged 50 and over, with little attention paid to its effects in healthy younger populations. The present research identified that increased BP may be beneficial to certain cognitive domains, while detrimental to others. In the young cohort (18–35 years), calculation skill showed a positive relationship with increasing diastolic blood

Table 3 Comparison between heart rate variability (active) and cognitive domain performance in three age groups.

Age group	Active HRV parameter	Cognitive domain	r	p
18-35 years	LF	Attention	-0.3	0.03
		Language	-0.31	0.03
36-50 years	HF	Naming	-0.35	0.03
		Repetition	-0.33	0.03
		Similarity	-0.37	0.02
	TP	Similarity	-0.32	0.04
51-65 years	LF	Orientation	-0.36	0.04
	LF:HF	Total score (Cognistat)	0.39	0.02
	TP	Orientation	-0.37	0.03

Key: HF=High frequency; LF=Low frequency; LF:HF=Sympathovagal balance; TP=Total power

pressure (DBP), seemingly contradicting findings from a retrospective cross-sectional study ($n = 5,077$) [28]. Judgement ability, on the other hand, a sub-skill of the reasoning domain [13], was shown to diminish with increasing DBP, agreeing with other current findings [28]. Increased systolic blood pressure (SBP) was found to improve performance in the memory, repetition and recall domains. This conflicts with a 2006 study by Wharton's laboratory that assessed BP and cognitive function in 18–21 year olds ($n = 105$) [52]. The group also reported a positive relationship between BP and visual search and spatial orientation tasks, whilst the present study showed increasing DBP impaired orientation performance.

The literature suggests that midlife hypertension increases the risk of dementia more than late-life hypertension [2]. Pathological changes (such as white matter lesions) noted in the brain during hypertensive states midlife reflect similar pathological changes seen in the early stages of Alzheimer's disease [40]. Current findings showed higher SBP and DBP were associated with better performance in similarity and calculation domains, but impaired comprehension, orientation, and attention domains. This is supported by the findings of a retrospective study ($n = 5,838$) [45]. Although, the recent Bogalusa Heart Study [21] presented results inconsistent with this association, reporting an inverse relationship between SBP and 63% of the cognitive domains of the neuropsychological assessments ($n = 351$). The vast majority of BP and cognition studies demonstrate SBP and DBP as predictors for cognitive impairment in older age. The present study found increased SBP and DBP linked to impaired recall ability in the older age group (51–65 years). This is supported by various studies, recall being a sub-skill of short-term memory [18]. The current study identified recall domain performance with high BP as a strong predictor for cognitive impairment. Increasing awareness of the importance of BP control in older age is a vital step towards manipulating modifiable risk factors to preserve cognitive function.

4.2 HRV and Cognitive Function (Baseline)

The current study showed LF was inversely linked to recall performance in the 18–35 year old group during baseline. Higher LF reflects sympathetic dominance, which is associated with increased BP, stress, cardiovascular disease and heightened mortality risk [7]. Construction skill decreased as HF and TP increased (baseline), reflecting parasympathetic dominance. This relationship may be influenced by unrefined construction abilities as visuospatial construction skills are still forming during adolescence and young adulthood [42]. Higher HF was linked with decreased ability in the naming domain. Prior to young adulthood there is a sharp improvement in the naming domain, as the frontal lobe is maturing, whereas only a gradual increase thereafter as one progresses towards midlife [6]. In older age, naming ability declines, not caused by a loss of vernacular, but rather increased difficulty in accessing existing knowledge stores, learning, and integrating new information. These associations reinforce that cardiovascular and neurocognitive systems do not operate in isolation from one another. Assessing the effects of autonomic activity over the lifetime may prompt adjustment of maintenance methods for autonomic control to prevent future onset of cognitive decline. Higher cardiac activity during middle age for LF and TP was associated with lower scores in the attention, construction, and repetition domains. Catecholamine exposure produced during the sympathetic (LF) response (particularly dopamine and norepinephrine) attenuates working memory function, as well as other higher order skills [39]. Research shows that imbalanced HRV during resting state heightens the risk of cognitive impairment, cardiovascular disease, vascular dementia and Alzheimer's disease [33]. Low HRV was linked to lowered orientation skill in older age (51–65 years) (baseline). Clinically, diminishing orientation function is one of the earliest predictors of cognitive decline seen in Alzheimer's disease [30]. Sympathetic dominance accompanying reduced cognitive function in old age has been supported by many studies [31, 41] and has been linked to other diseased states including hypertension and anxiety disorders [3, 26].

4.3 HRV and Cognitive Function (Active)

Measuring HRV changes during an active cognitive task is crucial as reactivity to a stimulus is an essential mechanism reflecting the body's ability to adapt to the surroundings and supply metabolic needs. Although there are few literature comparisons, findings from the present study showed that in the youngest age group higher LF was linked to lower language and attention ability during performance of the cognitive task. In contrast, increasing parasympathetic activity (HF and TP) diminished similarity, repetition and naming skill during midlife (36–50 years). Increased parasympathetic drive has been associated with lowered cerebral perfusion, potentially causing ischemic injury and impairing cognition which may progress into dementia [24]. Higher LF was inversely correlated to orientation ability yet positively linked to total Cognistat score in older age participants (51–65 years). Heightened sympathovagal balance was linked to better total Cognistat score in the older sample which suggests sympathetic drive benefits overall cognitive function in the 51–65 year group. This has also been supported by several other studies [20, 22, 29, 35].

5 Conclusion

This research highlighted the prospective use of autonomic markers such as HRV (LF, HF, TP and LF:HF) and BP in relation to physiological aging to be utilised as early biomarkers of cognitive impairment. Future studies within the Neuroscience Research Unit, University of Technology, Sydney see to analysing sex effects, increasing sample size and assessing a clinical sample. This will help identify when cognitive function is most susceptible to autonomic changes. Early detection of those at higher risk of cognitive impairment would allow for preventative measures (such as anti- hypertensive use and autonomic biofeedback) to be applied earlier in life, to prevent or delay abnormal cognitive decline and progression into dementia. This would not only result in economic and social benefits, but also reduce the burden on carers and ultimately and most importantly preserve cognition in our aging population.

Acknowledgements

We acknowledge the Neuroscience Research Unit (led by A/Prof Sara Lal), School of Medical and Molecular Biosciences, University of Technology Sydney (UTS) for support. We thank the Science Faculty, UTS for PhD scholarship support for the first author and Mitesh Patel for data pre-processing. We also thank the Alzheimer's Australia Dementia Research Foundation for providing top-up PhD scholarship support for the first author.

References

[1] Alzheimer's Disease International, World Alzheimer Report 2010: The global economic impact of dementia. Alzheimer's Disease International, pp. 1–10, 2010

[2] B. B. Bendlin, C. M. Carlsson, C. E. Gleason, S. C. Johnson, A. Sodhi, C. L. Gallagher et al. Midlife predictors of Alzheimer's disease. Maturitas, 65(2):131–137, 2010.

[3] G. Berntson and J. Cacioppo. Heart rate variability: Stress and psychiatric conditions. In M. Malik and A. Camm (Eds.), Dynamic Electrocardiography, pp. 56–63. New York, Blackwell/Futura. 2004.

[4] M. M. Budge, C. De Jager, E. Hogervorst, A. D. Smith, and A. The Oxford Project to investigate memory, total plasma homocysteine, age, systolic blood pressure, and cognitive performance in older people. Journal of the American Geriatrics Society, 50(12):2014–2018, 2002.

[5] J. Chen, K. Lin, and Y. Chen. Risk factors for dementia. Journal of the Formosan Medical Association, 108(10):754–764, 2009.

[6] F. I. M. Craik and E. Bialystok. Cognition through the lifespan: Mechanisms of change. Trends in Cognitive Sciences, 10(3):131–138, 2006.

[7] O. V. Crowley, P. S. McKinley, M. M. Burg, J. E. Schwartz, C. D. Ryff, M. Weinstein et al., The interactive effect of change in perceived stress and trait anxiety on vagal recovery from cognitive challenge. International Journal of Psychophysiology, 2011 (in press).

[8] C. de Carli. Mild cognitive impairment: Prevalence, prognosis, aetiology, and treatment. The Lancet Neurology, 2(1):15–21, 2003.

[9] L. De Leon. Brain cognitive function and heart rate variability: Sympathetic and parasympathetic associations. Honours Thesis, Faculty of Science, University of Technology, Sydney, 2009.

[10] A. Di Carlo, M. Baldereschi, L. Amaducci, S. Maggi, F. Grigoletto, G. Scarlato et al. Cognitive impairment without dementia in older people: Prevalence, vascular risk factors, impact on disability. The Italian Longitudinal Study on aging. Journal of the American Geriatrics Society, 48:775–782, 2000.

[11] J. Djernes. Prevalence and predictors of depression in populations of elderly: A review. Acta Psychiatrica Scandinavica, 113:372–387, 2006.

[12] D. L. Eckberg. Sympathovagal balance: A critical appraisal. Circulation, 96(9):3224–3232, 1997.

[13] C. Engelhart, N. Eisenstein, V. Johnson, J. Wolf, J. Williamson, D. Steitz et al., Factor structure of the Neurobehavioral Cognitive Status Exam (COGNISTAT) in

healthy, and psychiatrically and neurologically impaired, elderly adults. The Clinical Neuropsychologist, 13(1):109–111, 1999.

[14] M. Ewers, C. Walsh, J. Q. Trojanowski, L. M. Shaw, R. C. Petersen, C. R. Jack Jr. et al. Prediction of conversion from mild cognitive impairment to Alzheimer's disease dementia based upon biomarkers and neuropsychological test performance. Neurobiology of Aging, 33:1203–1214, 2012.

[15] D. Ewing, C. Martin, R. Young, and B. Clarke. The value of cardiovascular autonomic function tests: 10 years experience in diabetes. Diabetes Care, 8:491–498, 1985.

[16] M. F. Folstein, S. E. Folstein, and P. R. McHugh. Mini-Mental State Practical Method for grading cognitive status of patients for clinician. Journal of Psychiatric Research, 12:189–198, 1975.

[17] D. M. Geslani, M. C. Tierney, N. Herrmann, and J. P. Szalai. Mild cognitive impairment: An operational definition and its conversion rate to Alzheimer's disease. Dementia and Geriatric Cognitive Disorders, 19(5–6):383–389, 2005.

[18] P. J. Gianaros, P. J. Greer, C. M. Ryan, and J. R. Jennings. Higher blood pressure predicts lower regional grey matter volume: Consequences on short-term information processing. NeuroImage, 31(2):754–765, 2006.

[19] Z. Guo, M. Viitanen, L. Fratiglioni, and B. Winblad. Low blood pressure and dementia in elderly people: The Kungsholmen project. British Medical Journal, 312:805–808, 1996.

[20] Z. Guo, L. Fratiglioni, B. Winblad, and M. Viitanen. Blood pressure and performance on the Mini-Mental State Examination in the very old: Cross-sectional and longitudinal data from the Kungsholmen Project. American Journal of Epidemiology, 145:1106–1113, 1997.

[21] J. Gustat, J. Rice, B. Seltzer, and G. Berenson. Blood pressure and depression affect performance on cognitive tests in a biracial (black-white) middle age population: The Bogalusa heart study. Alzheimer's and Dementia, 7(4, Supplement):601–614, 2011.

[22] M. Kähönen-Väre, S. Brunni-Hakala, M. Lindroos, K. Pitkala, T. Strandberg, and R. Tilvis. Left ventricular hypertrophy and blood pressure as predictors of cognitive decline in old age. Aging Clinical and Experimental Research, 16:147–152, 2004.

[23] E. Kajantie and K. Räikkönen. Early life predictors of the physiological stress response later in life. Neuroscience & Biobehavioral Reviews, 35(1):23–32, 2010.

[24] S. P. Kennelly, B. A. Lawlor, and R. A. Kenny. Blood pressure and the risk for dementia: A double edged sword. Ageing Research Reviews, 8(2):61–70, 2009.

[25] E. A. Kensinger. Cognition in aging and age-related disease. In P. Hof and C. Mobbs (Eds.), Encyclopedia of Neuroscience, pp. 1055–1061. Academic Press, Oxford, 2009.

[26] D. Kim, L. Lipsitz, L. Ferrucci, R. Varadhan, J. Guralnik, M. Carlson et al. Associations between reduced heart rate variability and cognitive impairment in older disabled women in the community: Women's Health and Aging Study I. Journal of the American Geriatrics Society, 54(11):1751–1757, 2006.

[27] T. Koskinen, M. Kähönen, A. Jula, T. Laitinen, L. Keltikangas-Järvinen, J. Viikari et al. Short-term heart rate variability in healthy young adults: The Cardiovascular Risk in Young Finns Study. Autonomic Neuroscience, 145(1-2):81–88, 2009.

[28] M. B. Lande, J. M. Kaczorowski, P. Auinger, G. J. Schwartz, and M. Weitzman. Elevated blood pressure and decreased cognitive function among school-age children and adolescents in the United States. The Journal of Pediatrics, 143(6):720–724, 2003.

[29] L. Launer, K. Masaki, H. Petrovitch, D. Foley, and R. Havlik. The association between midlife blood pressure levels and late-life cognitive function: The Honolulu-Asia Aging Study. Journal of the American Medical Association, 274:1846–1851, 1995.

[30] A. Levey, J. Lah, F. Goldstein, K. Steenland, and D. Bliwise. Mild cognitive impairment: An opportunity to identify patients at high risk for progression to Alzheimer's disease. Clinical Therapeutics, 28(7):991–1001, 2006.

[31] M. Marin, C. Lord, J. Andrews, R. Juster, S. Sindi, G. Arsenault-Lapierre et al. Chronic stress, cognitive functioning and mental health. Neurobiology of Learning and Memory, 2011 (in press).

[32] K. Mathewson, J. Dywan, P. Snyder, W. Tays, and S. Segalowitz. Autonomic regulation and maze-learning performance in older and younger adults. Biological Psychology, 88(1):20–27, 2011.

[33] M. Morris, P. Scherr, L. Hebert, and D. Bennett. The cross-sectional association between blood pressure and Alzheimer's disease in a biracial community population of older persons. The Journals of Gerontology, 55(3):130–138, 2000.

[34] V. Napadow, R. Dhond, G. Conti, N. Makris, E. N. Brown, and R. Barbieri. Brain correlates of autonomic modulation: Combining heart rate variability with fMRI. NeuroImage, 42(1):169–177, 2008.

[35] R. Pandav, H. Dodge, S. DeKosky, and M. Ganguli. Blood pressure and cognitive impairment in India and the United States: A cross-national epidemiological study. Archives of Neurology, 60:1123–1128, 2003.

[36] M. Patel, S. K. L. Lal, D. Kavanagh, and P. Rossiter. Applying neural network analysis on heart rate variability data to assess driver fatigue. Expert Systems with Applications, 38(6):7235–7242, 2011.

[37] A. C. Phillips. Blunted cardiovascular reactivity relates to depression, obesity, and self-reported health. Biological Psychology, 86(2):106–113, 2011.

[38] J. Pumprla, K. Howorka, D. Groves, M. Chester, and J. Nolan. Functional assessment of heart rate variability: Physiological basis and practical applications. International Journal of Cardiology, 84(1):1–14, 2002.

[39] S. Qin, E. J. Hermans, H. J. F. van Marle, J. Luo, and G. Fernández. Acute psychological stress reduces working memory-related activity in the dorsolateral prefrontal cortex. Biological Psychiatry, 66(1):25–32, 2009.

[40] N. Raz and K. M. Rodrigue. Differential aging of the brain: Patterns, cognitive correlates and modifiers. Neuroscience & Biobehavioral Reviews, 30(6):730–748, 2006.

[41] N. Raz, K. M. Rodrigue, K. M. Kennedy, and J. D. Acker. Vascular health and longitudinal changes in brain and cognition in middle-aged and older adults. Neuropsychology, 21(2):149–157, 2007.

[42] K. Rubia, Z. Hyde, R. Halari, V. Giampietro, and A. Smith. Effects of age and sex on developmental neural networks of visual-spatial attention allocation. NeuroImage, 51(2):817–827, 2010.

[43] L. H. Schwamm, C. Van Dyke, R. J. Kiernan, E. L. Merrin, and J. Mueller. The Neurobehavioral Cognitive Status Examination: Comparison with the Cognitive Capacity Screening Examination and the Mini-Mental State Examination in a neurosurgical population. Annals of Internal Medicine, 107(4):486–491, 1987.

[44] A. Shehab and A. Abdulle. Cognitive and autonomic dysfunction measures in normal controls, white coat and borderline hypertension. BioMed Central-Cardiovascular Disorders, 11:11–13, 2011.

[45] A. Singh-Manoux and M. Marmot. High blood pressure was associated with cognitive function in middle age in the Whitehall II study. Journal of Clinical Epidemiology, 58(12):1308–1315, 2005.

[46] C. D. Smith, H. Chebrolu, D. R. Wekstein, F. A. Schmitt, and W. R. Markesbery. Age and gender effects on human brain anatomy: A voxel-based morphometric study in healthy elderly. Neurobiology of Aging, 28(7):1075–1087, 2007.

[47] J. A. Suhr, J. C. Stewart, and C. R. France. The relationship between blood pressure and cognitive performance in the Third National Health and Nutrition Examination Survey (NHANES III). Psychosomatic Medicine, 66(3):291–297, 2004.

[48] J. F. Thayer, J. J. Sollers III, D. M. Labiner, M. Weinand, A. M. Herring, R. D. Lane et al. Age-related differences in prefrontal control of heart rate in humans: A pharmacological blockade study. International Journal of Psychophysiology, 72(1):81–88, 2009.

[49] J. F. Thayer, S. S. Yamamoto, and J. F. Brosschot. The relationship of autonomic imbalance, heart rate variability and cardiovascular disease risk factors. International Journal of Cardiology, 141(2):122–131, 2010.

[50] J. F. Thayer, F. hs, M. Fredrikson, J.J. Sollers IIII, and T. D. Wager. A meta-analysis of heart rate variability and neuroimaging studies: Implications for heart rate variability as a marker of stress and health. Neuroscience and Biobehavioral Reviews, 36(2)747–756, 2010.

[51] S. Waldstein, P. Giggey, J. Thayer, and A. Zonderman. Nonlinear relations of blood pressure to cognitive function: The Baltimore Longitudinal Study of Aging. Hypertension, 45:374–379, 2005.

[52] W. Wharton, E. Hirshman, P. Merritt, B. Stangl, K. Scanlin, and L. Krieger. Lower blood pressure correlates with poorer performance on visuospatial attention tasks in younger individuals. Biological Psychology, 73(3):227–234, 2006.

Biographies

Louisa Giblin obtained her Bachelor's degree in Medical Science with first class honours at the University of Technology, Sydney. Louisa is currently completing a PhD within the Neuroscience Research Unit, School of Medical and Molecular Biosciences under the principal supervision of Professor Sara Lal. Louisa's research is focused on the links between heart rate variability and cognitive function.

Levin De Leon obtained his Bachelor's degree in Medical Science with first class honours at the University of Technology, Sydney under the principal supervision of Professor Sara Lal. He is now in his clinical years completing a Doctor of Medicine degree at the University of Melbourne with ambitions of specializing in neurosurgery with continual research in cognitive decline, particularly in Alzheimer's disease.

Lisa Smith obtained a Bachelor's degree in Medical Science with first class honours at the University of Technology, Sydney. Lisa currently studies dentistry at Griffith University, Queensland.

Tamara Sztynda completed a Bachelor of Science, Master of Science and PhD (Medicine) at the University of Melbourne. Dr. Sztynda is a Senior Lecturer at UTS within the School of Medical and Molecular Biosciences and the Program Director for the Bachelor of Forensic Biology in Biomedical Science. Dr. Sztynda is an Associate within the National Institute of Forensic Science, a member of the Australian and New Zealand Forensic Science Society and the NSW Histotechnology Group. Dr. Sztynda's research interests are in histopathology and forensic biology and currently she has research students who she is co-supervising with Professor Sara Lal in the forensic application of optical flow analysis for detection of facial emotions.

Sara Lal (PhD, MAppSc, BSc, GCHE, DipLaw) is a neuroscientist in the School of Medical and Molecular Biosciences at the University of Technology, Sydney. Professor Lal is the principal supervisor on this research publication. Professor Lal has supervised multiple research students and published widely in medical, scientific and engineering journals in areas of neuroscience, fatigue, algorithms, cognitive sciences, countermeasures and in cardiovascular research. She has attracted multiple competitive grants.

Online Manuscript Submission

The link for submission is: www.riverpublishers.com/journal

Authors and reviewers can easily set up an account and log in to submit or review papers.

Submission formats for manuscripts: LaTeX, Word, WordPerfect, RTF, TXT.
Submission formats for figures: EPS, TIFF, GIF, JPEG, PPT and Postscript.

LaTeX

For submission in LaTeX, River Publishers has developed a River stylefile, which can be downloaded from http://riverpublishers.com/river_publishers/authors.php

Guidelines for Manuscripts

Please use the Authors' Guidelines for the preparation of manuscripts, which can be downloaded from http://riverpublishers.com/river_publishers/authors.php

In case of difficulties while submitting or other inquiries, please get in touch with us by clicking CONTACT on the journal's site or sending an e-mail to: info@riverpublishers.com